ELEMENTARY ALGEBRA

2nd Edition

Joan Van Glabek, Ph.D.
Edison College
Naples, FL

Contributing Editor
John Niman, Ph.D.
Hunter College of the City University of New York
New York, NY

Collins
An Imprint of HarperCollins*Publishers*

An American BookWorks Corporation Production

HarperCollins books may be purchased for educational, business, or sales promotional use. For information please write: Special Markets Department, HarperCollins Publishers, 10 East 53rd Street, New York, NY 10022.

Library of Congress Cataloging-in-Publication Data

Van Glabek, Joan.
Elementary algebra / Joan Van Glabek ; contributing editor, John Niman. – 2nd ed.
p. cm.
ISBN: 978-0-06-088148-1
ISBN-10: 0-06-088148-8
1. Algebra—Problems, exercises, etc. I. Niman, John. II. Title.

QA157.V355 2006
512–dc22 2006051734

---

06 07 08 09 10 CW 10 9 8 7 6 5 4 3 2 1

Contents

Preface

This book presents the material for a firm foundation in elementary algebra. There are many worked examples with enough steps to help the beginning algebra student follow the work from one step to the next step. As you progress through the material, if you are able to skip a step or two, great! A set of exercises and answers appear at the end of each chapter to allow the reader to practice and receive immediate feedback. Keep pencil and paper handy—reading and working through this book will give you the skills necessary to continue your study of mathematics.

Joan Van Glabek, Ph.D.

CHAPTER 1

Number Systems and the Basics

This chapter contains a review of arithmetic topics and extends these concepts to algebra. You may be able to skip part (or all) of this chapter if your mathematics background is current. Consider working the problems at the end of the chapter as a pretest. If you can work the problems at the end of the chapter correctly, move ahead to Chapter 2.

1.1 NOTATION AND ORDER OF OPERATIONS

The four basic operations used in arithmetic—addition, subtraction, multiplication, and division—are also used in algebra. This section reviews some language and symbols from arithmetic and describes the uses of some of these familiar symbols in algebra. You may want to highlight words that are unfamiliar and refer back to this section as necessary. Variables (letters like x and y) will be used in some of the exercises to represent numbers.

THE BASIC OPERATIONS

The following table reviews some of the language and symbols associated with addition, subtraction, multiplication, and division:

Operation	Symbol	Example	English Phrases
Addition	$+$	$4 + 6$	The sum of 4 and 6 4 plus 6 4 added to 6
Subtraction	$-$	$5 - 2$	The difference of 5 and 2 5 minus 2 5 less 2 2 subtracted from 5
Multiplication	$\cdot$	$3 \cdot 2$	The product of 3 and 2 3 times 2
Division	$\div$	$6 \div 2$	The quotient of 6 and 2 6 divided by 2 2 into 6

We avoid the use of the symbol "×" for multiplication in algebra to prevent confusion between the variable x and the symbol "×" to imply multiplication. Other ways to show multiplication in algebra may involve the use of parentheses.

Multiplication with Numbers	Multiplication with Variables
3(4)	$a(b)$
(3)4	$(a)b$
(3)(4)	$(a)(b)$
No numerical equivalent	ab

Multiplication is implied in the following cases: when there is no symbol between two variables; when there is no symbol between a variable and a number; and when there is no symbol between a variable and a parenthesis. You cannot use this convention when multiplying two numbers, however. 34 means thirty-four, never 3 times 4.

Likewise, there are several ways to imply division.

Division with Numbers	Division with Variables
$6 \div 2$	$a \div b$
$\frac{a}{b}$	$\frac{a}{b}$
$2\overline{)6}$	$b\overline{)a}$
6/2	a/b

SIMPLIFYING

In arithmetic, many problems contain only one of the four basic operations of addition, subtraction, multiplication, and division. In algebra, there will often be *more* than one operation involved. Thus, you will often see directions such as "simplify," which means you should perform any indicated operations.

GROUPING SYMBOLS

The following symbols are used to group numbers together:

Symbols	Name of the Symbols
( )	Parentheses
[ ]	Brackets
{ }	Braces
||	Absolute value bars
___	Fraction bar

When simplifying, perform operations inside parentheses, brackets, braces, and absolute value bars first. The fraction bar as a grouping symbol means to simplify above and below the bar separately.

Exercise 1.1

Simplify:

a) $(4 + 8)$

b) $[10 - 5]$

c) $\dfrac{4+5}{6-3}$

d) $(7 - 3) + (6 - 4)$

Solution 1.1

a) $(4 + 8) = (12) = 12$ — Add inside the parentheses.

b) $[10 - 5] = [5] = 5$ — Subtract inside the brackets.

c) $\dfrac{4+5}{6-3} = \dfrac{9}{3}$ — $\dfrac{\text{Simplify above the bar.}}{\text{Simplify below the bar.}}$

$= 3$ — Reduce the fraction.

d) $(7 - 3) + (6 - 4) = (4) + (2)$ — Simplify within each set of parentheses.

$= 6$ — Simplify by adding.

LANGUAGE

The numbers and/or variables involved in the four basic operations have names, as given in the following table:

Example	Terms
$5 + 8 = 13$	5 and 8 are called **addends.** 13 is called the **sum.**
$10 - 1 = 9$	10 is called the **minuend.** 1 is called the **subtrahend.** 9 is called the **difference.**
$2(7) = 14$	2 and 7 are called **factors.** 14 is called the **product.**
$\frac{48}{4} = 12$	48 is called the **numerator.** 4 is called the **denominator.** 12 is called the **quotient.**
$\begin{array}{r} 3 \\ 6\overline{)19} \\ \underline{18} \\ 1 \end{array}$	19 is called the **dividend.** 6 is called the **divisor.** 3 is called the **quotient.** 1 is called the **remainder.**

Exercise 1.2

For each sentence, write an equivalent expression in symbols:

a) The sum of x and 2

b) The difference of 5 and y

c) The product of 4 and a

d) The quotient of z and 5

Solution 1.2

a) $x + 2$ Sum means addition.

b) $5 - y$ Difference means subtraction.

c) $4 \cdot a$ Product means multiplication.

d) $\frac{z}{5}$ Quotient means division.

EXPONENTS

We use exponents to write repeated multiplications.

"2" is the exponent (also called the power).

↓

$5^2 = 5 \cdot 5 = 25$

↑

"5" is the base.

The exponent (or power) tells you how many times to use the base as a factor.

Exponent Form	How to Read It	Meaning
3^2	3 squared or 3 to the second power	$3 \cdot 3 = 9$
4^3	4 cubed or 4 to the third power	$4 \cdot 4 \cdot 4 = 64$
2^4	2 to the fourth power	$2 \cdot 2 \cdot 2 \cdot 2 = 16$

Exercise 1.3

Expand and multiply:

a) 2^5

b) 5^2

c) 10^3

Solution 1.3

a) $2^5 = 2 \cdot 2 \cdot 2 \cdot 2 \cdot 2 = 32$ — 2 is the base, 5 is the exponent.

b) $5^2 = 5 \cdot 5 = 25$ — 5 is the base, 2 is the exponent.

c) $10^3 = 10 \cdot 10 \cdot 10 = 1{,}000$ — 10 is the base, 3 is the exponent.

ORDER OF OPERATIONS

Often, there will be more than one operation or one set of grouping symbols in a problem. Mathematicians have agreed upon the order of operations to be followed in solving problems.

1. Simplify inside grouping symbols, starting with the innermost pair and working out.
2. Simplify exponents.
3. Perform multiplication and division steps in order from left to right.
4. Perform addition and subtraction steps in order from left to right.

You may find it helpful in remembering the order of operations to memorize the expression: P E M D A S or Please Excuse My Dear Aunt Sally, which stands for Parentheses, Exponents, Multiplication, Division, Addition, Subtraction.

Exercise 1.4

Simplify $\{5 + [8 - (3 - 1)]\}$.

Solution 1.4

$\{5 + [8 - (3 - 1)]\} = \{5 + [8 - (2)]\}$ — Work inside the parentheses.

$= \{5 + [6]\}$ — Work inside the brackets.

$= \{11\}$ — Work inside the braces.

$= 11$ — This is the answer.

Exercise 1.5

Simplify $4 + 6 \div 2 + 3 \cdot 5 - 4$.

Solution 1.5

$4 + 6 \div 2 + 3 \cdot 5 - 4 = 4 + 3 + 3 \cdot 5 - 4$	Multiply and divide in order from left to right: $6 \div 2 = 3$.
$= 4 + 3 + 15 - 4$	Multiplication is next: $3 \cdot 5 = 15$.
$= 18$	Add and subtract in order from left to right.

Exercise 1.6

Simplify $2^3 \div 2 + 4(9 - 2^2)$.

Solution 1.6

$2^3 \div 2 + 4(9 - 2^2) = 2^3 \div 2 + 4(9 - 4)$	Work inside the parentheses. Simplify exponents.
$= 2^3 \div 2 + 4(5)$	Work inside the parentheses.
$= 8 \div 2 + 4(5)$	Simplify exponents: $2^3 = 8$.
$= 4 + 4(5)$	Divide: $8 \div 2 = 4$.
$= 4 + 20$	Multiply: $4(5) = 20$.
$= 24$	Add.

Exercise 1.7

Simplify $\frac{2(6-3)+8}{4(2)-6}$.

Solution 1.7

Remember, a fraction bar is a grouping symbol. Simplify above the fraction bar and below the fraction bar separately.

$\frac{2(6-3)+8}{4(2)-6} = \frac{2(3)+8}{4(2)-6}$	Work inside the parentheses.

$\dfrac{6+8}{8-6}$ Multiply.

$= \dfrac{14}{2}$ Add in the numerator, subtract in the denominator.

$= 7$ Simplify.

1.2 NUMBER LINES AND SIGNED NUMBERS

You can use a **number line** to picture the real numbers. Begin by drawing a line and labeling a convenient point as 0. Mark a point to the right of 0 and label it as 1. The distance between 0 and 1 gives you a way to measure and label other numbers on the number line.

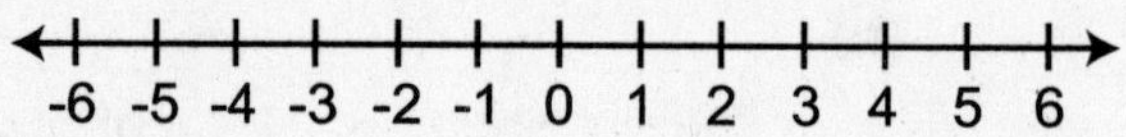

Notice that the positive numbers are located to the right of zero on the number line, and the negative numbers are located to the left of zero. Numbers increase in value from the left to the right of the number line. Note also that a negative number must have a negative sign preceding it, while a positive number can be written either with a positive sign preceding it or with no sign preceding it. Because fractions and decimals are real numbers, they also can be pictured (or graphed) on the number line.

Exercise 1.8

Label points on a number line for the following numbers:

$2, 5, -1, -4, \frac{1}{2}, 3.8$

Solution 1.8

Use a point on the number line to locate each point:

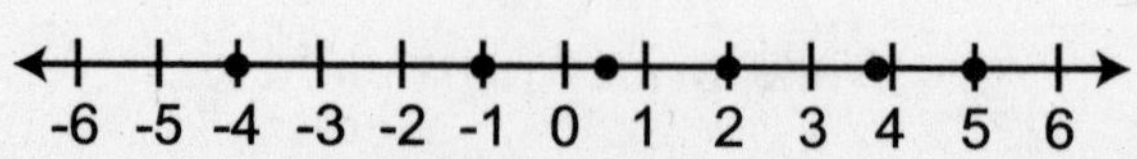

EQUALITY AND INEQUALITY SYMBOLS

You can use a number line to decide the order relationship between numbers. You will need to use the following symbols:

Symbol	English Translation	Example
$<$	Less than	$2 < 4$
$>$	Greater than	$5 > -1$
$\leq$	Less than or equal to	$8 \leq 10$
$\geq$	Greater than or equal to	$4 \geq 4$
$=$	Equal to	$-2 = -2$
$\neq$	Not equal to	$5 \neq 3$

Exercise 1.9

Write an equivalent expression in symbols:

a) The sum of 4 and x is greater than 12.

b) The product of 5 and y is less than or equal to 20.

c) The product of 3 and x is equal to the difference of x and 6.

Solution 1.9

a) $4 + x > 12$ Sum means addition.

b) $5y \leq 20$ Product means multiplication.

c) $3x = x - 6$ Product means multiplication; difference means subtraction.

When a number a is to the left of a number b on the number line, a is less than b. If c is to the right of b on the number line, c is greater than b. Consider the following number line and the inequalities it represents:

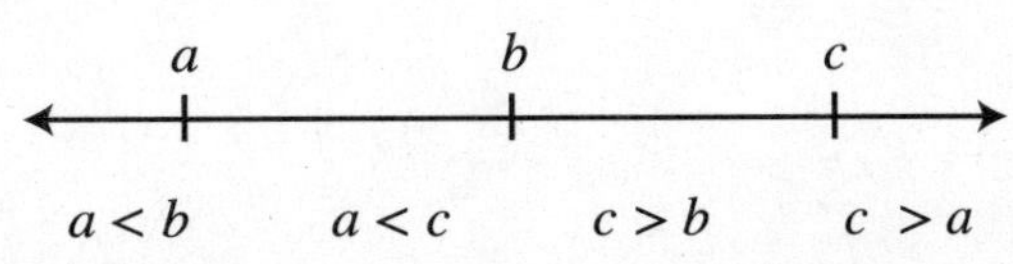

You may change the order in which an inequality is written as long as you keep the symbol "pointing" at the same number. Thus, an inequality stated as $a < b$ can be written as $b > a$.

Although it is usually easy to compare two positive numbers, you may want to draw a number line to help you compare negative numbers.

Exercise 1.10

Insert "<" or ">" to make each statement true. Use a number line if you need help:

a) -1 __ 4

b) -2 __ 4

c) $-3\frac{1}{2}$ __ 3

Solution 1.10

a) $-1 < 4$, because -1 is to the left of 4.

b) $-2 > -4$, because -2 is to the right of -4.

c) You may want to draw a number line to help with this one.

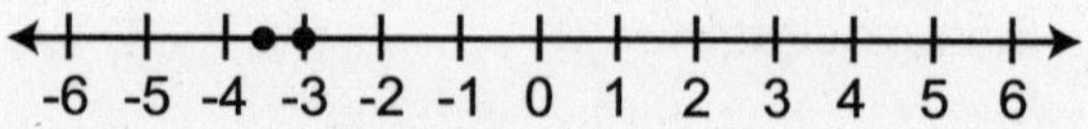

Notice that $-3\frac{1}{2}$ is to the left of -3 and, therefore, $-3\frac{1}{2} < -3$.

ABSOLUTE VALUE

You use a number line to define absolute value. The distance of a number a from zero is its absolute value, written $|a|$.

Exercise 1.11

Find the absolute value of:

a) 2

b) -3

c) 0

Solution 1.11

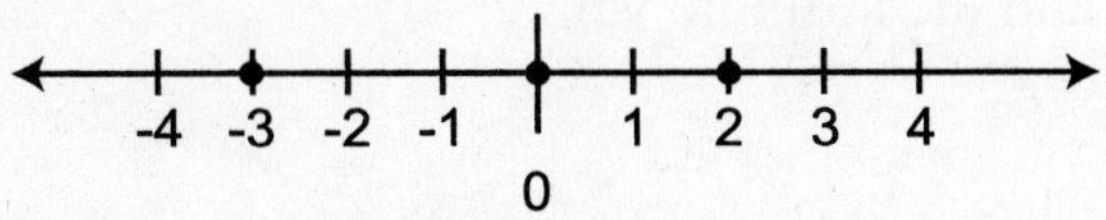

You can use the number line to note that

a) |2| = 2, because 2 is 2 units away from 0.

b) |–3| = 3, because –3 is 3 units away from 0.

c) |0| = 0, because 0 is 0 units away from 0.

OPPOSITE OR ADDITIVE INVERSE

You also use a number line to describe the **additive inverse** or **opposite** of a number. The **opposite** of a number a is the number that is the same distance from 0 as a, but on the opposite side of 0.

Exercise 1.12

Find the opposite of:

a) 3

b) –2

c) 0

Solution 1.12

a) Using a number line, you can see that –3 is the opposite of 3.

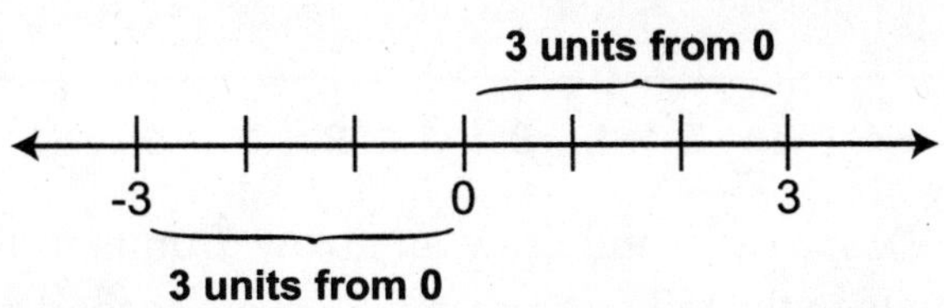

b) 2 is the opposite of –2.

c) 0 is the only number that is its own opposite.

Another way to write the opposite of a number is to use the symbol "–". For example, the opposite of 3 is written –3. The opposite of –2 is written – (–2). Notice from Exercise 1.12 that – (–2) = 2.

Exercise 1.13

For each of the given numbers, state the opposite and the absolute value :

a) 6

b) –5

c) x

Solution 1.13

a) $-(6) = -6$ The opposite of 6 is –6.

$|6| = 6$ The absolute value of 6 is 6.

b) $-(-5) = 5$ The opposite of –5 is 5.

$|-5| = 5$ The absolute value of –5 is 5.

c) $-(x) = -x$ The opposite of x is $-x$.

$|x| = |x|$ The absolute value of x is the absolute value of x. We cannot drop the absolute value bars because we don't know whether x represents a positive or negative number.

1.3 ADDING REAL NUMBERS

Addition of two signed numbers can be demonstrated using a number line. For example, 2 + 4 would be shown by starting at 0, drawing an arrow 2 units to the right, and then drawing an arrow 4 more units to the right. Where this second arrow ends, 6 is the sum.

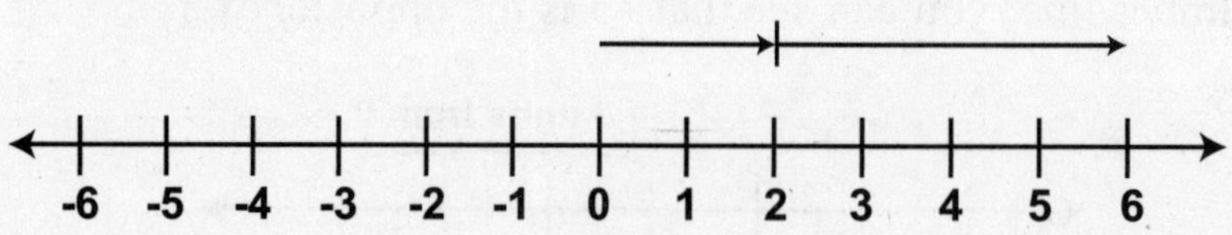

Similarly, to add (–2) + (–3), start at 0 and draw an arrow 2 units to the left to represent –2. Then draw an arrow 3 more units to the left to represent –3. This second arrow ends at –5, which is the sum.

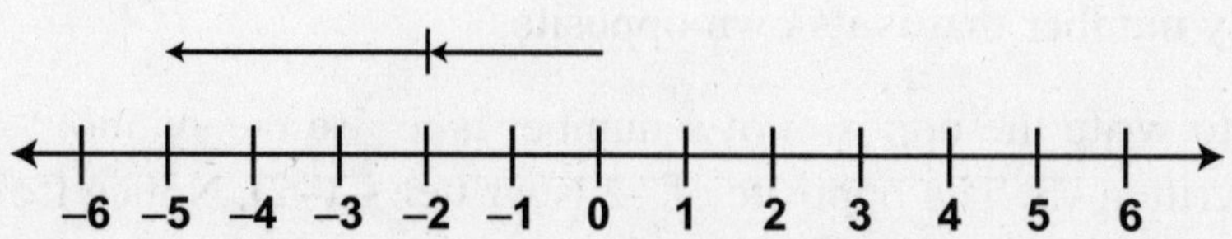

These two examples lead to the following rule for adding numbers with the same signs:

1. Add the absolute values of the numbers.
2. Write the common sign.

Exercise 1.14

Find the sums:

a) $5 + 2$

b) $-4 + (-7)$

Solution 1.14

a) $5 + 2 = 7$ Same signs of +.

b) $-4 + (-7)$

$4 + 7$ $|-4| = 4, |-7| = 7$

11 Add absolute values.

$-4 + (-7) = -11$ Write the common sign –.

A number line can also be used to demonstrate adding numbers with different signs. For example, add $5 + (-3)$ by beginning with an arrow from 0 to the right 5 units. Then draw an arrow from the end of that arrow 3 units to the left. The second arrow ends at the sum, which is 2.

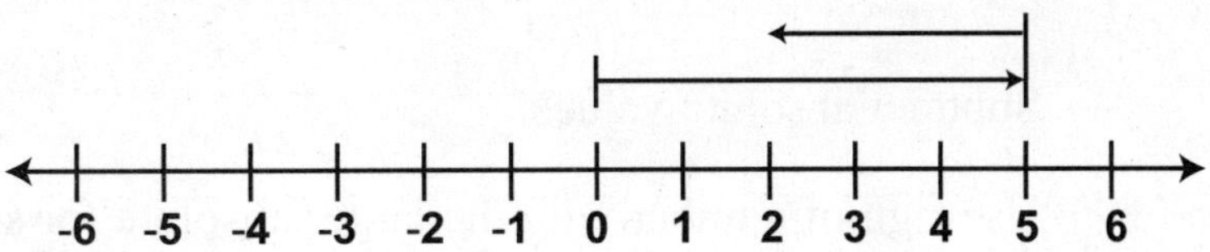

You can also show $-6 + 2$ on a number line. Begin at 0 and draw an arrow 6 units to the left. Then draw an arrow from the end of that arrow 2 units to the right. The second arrow ends at the sum, -4.

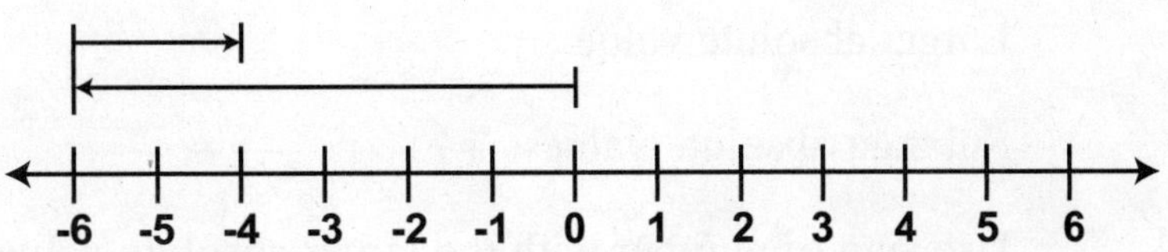

To add numbers with different signs without using a number line, do the following:

1. Find the absolute value of each number.
2. Subtract the number with the smaller absolute value from the number with the larger absolute value.
3. Write the sign of the number with the larger absolute value in front of the sum.

You may need to work several more exercises to convince yourself that the preceding rules work. (Remember that positive signs do not have to be written, but negative signs must be written.)

Exercise 1.15

Find the sums:

a) $4 + (-2)$

b) $3 + (-7)$

c) $(-6) + 9$

d) $(-5) + 3$

Solution 1.15

a) $4 + (-2)$

$\lvert 4\rvert = 4$	Larger absolute value.
$\lvert -2\rvert = 2$	
$4 - 2 = 2$	Subtract absolute values.
$4 + (-2) = 2$	Use sign of number with the larger absolute for sign of sum.

b) $3 + (-7)$

$\lvert 3\rvert = 3$	
$\lvert -7\rvert = 7$	Larger absolute value.
$7 - 3 = 4$	Subtract absolute values.
$3 + (-7) = -4$	Use sign of number with the larger absolute value.

c) $(-6) + 9$

$|-6| = 6$

$|9| = 9$ Larger absolute value.

$9 - 6 = 3$ Subtract.

$(-6) + 9 = 3$ Use sign of number with the larger absolute value.

d) $(-5) + 3$

$|-5| = 5$ Larger absolute value.

$|3| = 3$

$5 - 3 = 2$ Subtract.

$(-5) + 3 = -2$ Use sign of number with the larger absolute value.

Exercise 1.16

Find the sums:

a) $-6 + 4 + (-8)$

b) $(-7 + 12) + [(-2) + (-6)]$

c) $-8 + 2(-4 + 7) + 6$

Solution 1.16

a) $-6 + 4 + (-8)$

$= -2 + (-8)$ Add from left to right.

$= -10$ Same signs: Add and keep the sign.

b) $(-7 + 12) + [(-2) + (-6)]$

$= (5) + [-8]$ Work inside the parentheses; work inside the brackets.

$= -3$ Different signs: Use sign of the number with the larger absolute value.

c) $-8 + 2\,(-4 + 7) + 6$

$= -8 + 2\,(3) + 6$ Work inside the parentheses first.

$= -8 + 6 + 6$ Multiply.

$= -2 + 6$	Add from left to right.
$= 4$	Different signs: Use sign of number with larger absolute value.

1.4 SUBTRACTING REAL NUMBERS

Subtraction is actually defined in terms of addition. For example, 5 – 3 is defined to be 5 + (–3), where –3 represents the opposite of 3. This definition leads to the following rules for subtracting:

1. Change the subtraction symbol to addition.
2. Write the opposite of the number being subtracted.
3. Use the rules for adding signed numbers.

Exercise 1.17

Find the differences:

a) $6 - 11$

b) $-3 - 7$

c) $8 - (-3)$

d) $-12 - (-5)$

Solution 1.17

a) $6 - 11 = 6 + (-11)$	Change the subtraction symbol to +; the opposite of 11 is –11.
$= -5$	Different signs: Subtract and use sign of –11.
b) $-3 - 7 = -3 + (-7)$	Change the subtraction symbol to +; the opposite of 7 is –7.
$= -10$	Same signs: Add and use the common sign.
c) $8 - (-3) = 8 + (+3)$	Change the subtraction symbol to +; the opposite of –3 is 3.
$= 11$	Same signs: Add and use the common sign.
d) $-12 - (-5) = -12 + (+5)$	Change the subtraction symbol to +; the opposite of –5 is +5.
$= -7$	Different signs: Subtract and use sign of –12.

The following exercises are subtraction problems with more than two numbers. Some exercises involve grouping symbols and order of operation rules. Remember to work inside the grouping symbols first.

Exercise 1.18

Simplify each of the following:

a) $2 - 6 - 5$

b) $2 - (6 - 5)$

c) $7 - (-4 + 2)$

d) $(-5 - 4) - (-7 - 1)$

Solution 1.18

a) $2 - 6 - 5 = 2 + (-6) + (-5)$ — Change the subtraction symbol to +; the opposite of 6 is –6; the opposite of 5 is –5.

$= -4 + (-5)$ — Add from left to right.

$= -9$ — Same signs: Add and use the common sign.

b) $2 - (6 - 5) = 2 - (6 + (-5))$ — Work inside the parentheses first.

$= 2 - (1)$ — Different signs: Subtract and use the sign of 6.

$= 2 + (-1)$ — Change the subtraction symbol to +; the opposite of 1 is –1.

$= 1$ — Different signs: Subtract and use the sign of 2.

Note that exercises (a) and (b) differ by a set of parentheses, which lead to different answers. If there are no grouping symbols, you must work in order from left to right. If there are grouping symbols, you must work inside the grouping symbols first.

c) $7 - (-4 + 2) = 7 - (-2)$ — Work inside the parentheses first.

$= 7 + (2)$ — Change the subtraction symbol to +; the opposite of –2 is 2.

$= 9$ — Same signs: Add and use the common sign.

d) $(-5-4)-(-7-1)$

$= (-5+(-4))-(-7+(-1))$	Change the subtraction symbol to +; the opposite of 4 is –4; the opposite of 1 is –1.
$= (-9)-(-8)$	Work inside the parentheses.
$= (-9)+(8)$	Change the subtraction symbol to +; the opposite of –8 is 8.
$= -1$	Different signs: Subtract and use the sign of –9.

Exercise 1.19

Use the rules for order of operations to simplify each expression as much as possible:

a) $4 \cdot 5 - 2 \cdot 8$

b) $3 \cdot 5 - 2 \cdot 7 - 5 \cdot 4$

c) $2 \cdot 3^2 - 3 \cdot 2^3$

d) $2 \cdot 7^2 - 3 \cdot 4^2 - 4 \cdot 2^2$

Solution 1.19

a) $4 \cdot 5 - 2 \cdot 8$

$= 20 - 16$	Perform multiplications first.
$= 20 + (-16)$	Change the subtraction symbol to +; the opposite of 16 is –16.
$= 4$	Different signs: Subtract and use the sign of 20.

b) $3 \cdot 5 - 2 \cdot 7 - 5 \cdot 4$

$= 15 - 14 - 20$	Perform multiplications first.
$= 15 + (-14) + (-20)$	Change the subtraction symbol to +; the opposite of 14 is –14, opposite of 20 is –20.
$= 1 + (-20)$	Add from left to right. Different signs: Subtract and use the sign of 15.
$= -19$	Different sign: Subtract and use the sign of –20.

c) $2 \cdot 3^2 - 3 \cdot 2^3$

$= 2 \cdot 9 - 3 \cdot 8$	Simplify exponents first.
$= 18 - 24$	Perform multiplications next.
$= 18 + (-24)$	Change the subtraction symbol to +; the opposite of 24 is –24.
$= -6$	Different signs: Subtract and use the sign of –24.

d) $2 \cdot 7^2 - 3 \cdot 4^2 - 4 \cdot 2^2$

$= 2 \cdot 49 - 3 \cdot 16 - 4 \cdot 4$	Simplify exponents first.
$= 98 - 48 - 16$	Perform multiplications next.
$= 98 + (-48) + (-16)$	Change the subtraction symbol to +; the opposite of 48 is –48, the opposite of 16 is –16.
$= 50 + (-16)$	Add from left to right. Different signs: Subtract and use the sign of 98.
$= 34$	Different signs: Subtract and use the sign of 50.

TRANSLATING INTO SYMBOLS

Some subtraction phrases translate in a straightforward manner. For example, the "difference of 5 and x" is written $5 - x$. However, the phrase "subtract 3 from x" means $x - 3$.

Exercise 1.20

Write each phrase as an equivalent expression in symbols. Simplify, if possible:

a) The difference of 8 and 5

b) The difference of 5 and 8

c) Subtract 4 from –6.

d) Subtract 8 from the sum of x and 2.

Solution 1.20

a) $8 - 5 = 3$	Difference translates in order.
b) $5 - 8 = 5 + (-8) = -3$	Change the subtraction symbol to +; the opposite of 8 is –8.

c) $-6 - 4 = -6 + (-4) = -10$ Be careful with the phrase "subtract from."

d) $(x+2)-8=x+2+(-8)=x-6$ The sum of x and 2 is written in parentheses; simplify.

1.5 PROPERTIES OF REAL NUMBERS

The properties of real numbers are rules that hold true for *all* real numbers (including rational and irrational numbers). The following is a list of these properties using variables, along with an example of each property:

For real numbers *a, b,* and *c*:

Name	Property	Example	Look For
Commutative property of addition	$a + b = b + a$	$3 + 4 = 4 + 3$	Change in the order
Commutative property of multiplication	$a \cdot b = b \cdot a$	$4 \cdot 6 = 6 \cdot 4$	Change in the order
Associative property of addition	$(a + b) + c = a + (b + c)$	$(2 + 3) + 5 = 2 + (3 + 5)$	Grouping symbols around a different pair of numbers
Associative property of multiplication	$(a \cdot b) c = a (b \cdot c)$	$(2 \cdot 3) \cdot 6 = 2 \cdot (3 \cdot 6)$	Grouping symbols around a different pair of numbers
Additive identity property	$a + 0 = a$ and $0 + a = a$	$5 + 0 = 5$ and $0 + 5 = 5$	Adding 0 to a number
Multiplicative identity property	$a(1) = a$ and $1(a) = a$	$6(1) = 6$ and $1(6) = 6$	Multiplying a number times 1
Additive inverse property	$a + (-a) = 0$	$3 + (-3) = 0$	Adding a number and its opposite

Multiplicative inverse property	$a \cdot \frac{1}{a} = \frac{1}{a} \cdot a = 1$	$7 \cdot \frac{1}{7} = \frac{1}{7} \cdot 7 = 1$	Multiplying a number times 1 over the number
Distributive property	$a\,(b + c) = a\,(b) + a\,(c)$ and $(b + c)\,a = b\,(a) + c\,(a)$	$2\,(3 + x) = 2\,(3) + 2\,(x)$	Multiplying across a sum in parentheses

COMMUTATIVE AND ASSOCIATIVE PROPERTIES

Notice from the table above that the commutative property involves a change in the order of the numbers and variables while the associative property involves a change in the grouping of the numbers and variables. The following exercise demonstrates the use of both properties:

Exercise 1.21

State the property (or properties) that justifies each statement:

a) $-5 + 8 = 8 + (-5)$

b) $x2 = 2x$

c) $5 + (7 + 10) = (5 + 7) + 10$

d) $6(5y) = (6 \cdot 5)y$

e) $4 + (5 + a) = (5 + a) + 4$

Solution 1.21

a) $-5 + 8 = 8 + (-5)$ — Commutative property of addition; the order of the numbers changed, and addition was the operation.

b) $x2 = 2x$ — Commutative property of multiplication; the order of the variable and number changed, and multiplication was the implied operation.

c) $5 + (7 + 10) = (5 + 7) + 10$ — Associative property of addition; the order of the numbers 5, 7, and 10 stayed the same on both sides of the equal sign. The numbers within the parentheses changed from 7 and 10 on one side, and 5 and 7 on the other side.

d) $6(5y) = (6 \cdot 5)y$ — Associative property of multiplication; the order of the factors 6, 5, and y stayed the same on both sides of the equal sign. The parentheses grouped 5 and y on one side, and 6 and 5 on the other side.

e) $4 + (5 + a) = (5 + a) + 4$ — Commutative property of addition; notice that only the order changed, not the numbers and variables within the parentheses.

IDENTITY AND INVERSE PROPERTIES

0 is called the **additive identity**; 1 is called the **multiplicative identity**. The numbers a and $-a$ are called **additive inverses**. The numbers a and $\frac{1}{a}$, where $a \neq 0$, are called **multiplicative inverses**.

Exercise 1.22

State the property (or properties) that justifies each statement:

a) $6 + 0 = 6$

b) $1(p) = p$

c) $x + (-x) = 0$

d) $10\left(\frac{1}{10}\right) = 1$

Solution 1.22

a) $6 + 0 = 6$ — Additive identity property; 0 added to any number equals the original number.

b) $1(p) = p$ — Multiplicative identity property; 1 multiplied times any number equals the original number.

c) $x + (-x) = 0$ — Additive inverse property; x and $-x$ are called additive inverses, and they sum to 0.

d) $10\left(\frac{1}{10}\right) = 1$ Multiplicative inverse property; 10 and $\frac{1}{10}$ are called multiplicative inverses, and their product is 1.

DISTRIBUTIVE PROPERTY

The distributive property involves two operations, multiplication and addition. Be careful to use the number or variable in front of the parentheses to multiply times each term inside the parentheses.

Exercise 1.23

Apply the distributive property to rewrite each of the following:

a) $5(2 + 6)$

b) $4(x + y)$

c) $-2(2a + 3b)$

Solution 1.23

a) $5(2 + 6) = 5 \cdot 2 + 5 \cdot 6$ Apply the distributive property.

$= 10 + 30$ Perform multiplication before addition.

$= 40$

b) $4(x + y) = 4x + 4y$

c) $-2(2a + 3b) = (-2)2a + (-2)3b$

$= -4a + (-6b)$ You can use the associative property here.

$(-2)(2a) = (-2 \cdot 2)a = -4a$

$= -4a - 6b$ Usual form for the answer.

Before we introduced the distributive property, we told you to work inside the parentheses first. Note from part (a) in Exercise 1.23 that

$5(2 + 6) = 5(8) = 40$

and

$5 \cdot 2 + 5 \cdot 6 = 10 + 30 = 40$

The distributive property guarantees that you will get the same answer whether you simplify inside the parentheses first or distribute first. We will use this property extensively when solving equations in Chapter 2. The distributive property also allows you to simplify expressions such as $-(5 + 8)$ and $-(4c - 2d)$, in which case the negative symbol directly preceding the parentheses tells you to multiply the expression in parentheses by -1.

Exercise 1.24

Apply the distributive property to each of the following. Simplify, when possible:

a) $-(5 + 8)$

b) $-(4c - 2d)$

Solution 1.24

a) $-(5 + 8) = -1(5 + 8)$

$= -1(5) + -1(8)$ Apply the distributive property.

$= -5 + -8$ Perform multiplication before addition.

$= -13$

b) $-(4c - 2d) = -1(4c - 2d)$ A negative symbol directly preceding the parentheses tells you to multiply by -1.

$= -1(4c + (-2d))$ Use rules for subtraction.

$= -1(4c) + -1(-2d)$ Use the distributive property.

$= -4c + 2d$ Use rules for multiplying signed numbers.

COMBINING TWO OR MORE PROPERTIES

Some problems require the use of two or more properties. Refer to the table at the beginning of this section to help you decide which ones to look for in these exercises.

Exercise 1.25

State the property (or properties) that justifies each statement:

a) $2 + (8 + 5) = (8 + 2) + 5$

b) $x(6y) = 6(xy)$

Solution 1.25

a) $2 + (8 + 5) = (2 + 8) + 5$ Associative property of addition

$= (8 + 2) + 5$ Commutative property of addition

Note that first the parentheses were changed to contain 2 and 8. Then the order of the 2 and 8 was changed to $8 + 2$.

b) $x(6y) = (x6)y$ Associative property of multiplication

$= (6x)y$ Commutative property of multiplication

$= 6(xy)$ Associative property of multiplication

Note that first the parentheses were changed to contain x and 6. Then the order of the x and 6 was changed to $6x$. Finally, the parentheses were changed to contain x and y.

1.6 MULTIPLYING REAL NUMBERS

We now extend the arithmetic concept of multiplying positive integers to multiplication of real numbers. We begin with the rules for multiplying signed numbers.

MULTIPLYING TWO SIGNED NUMBERS

There are four cases to consider when multiplying two signed numbers:

1. A positive number times a positive number equals a positive number.
2. A negative number times a negative number equals a positive number.
3. A positive number times a negative number equals a negative number.
4. A negative number times a positive number equals a negative number.

Many students remember only these two rules when multiplying two numbers:

* Same signs = positive product
* Different signs = negative product

Exercise 1.26

Multiply:

a) 4(6)

b) $(-2)(-5)$

c) $3(-6)$

d) $(-2)(8)$

Solution 1.26

a) $4(6) = 24$ Same signs: positive product.

b) $(-2)(-5) = 10$ Same signs: positive product.

c) $3(-6) = -18$ Different signs: negative product.

d) $(-2)(8) = -16$ Different signs: negative product.

MULTIPLYING MORE THAN TWO NUMBERS

When two or more factors are involved in the multiplication, you may find the following shortcuts useful:

* **Even and odd number of factors :** If there are an even number of negative factors, the product is positive; if there are an odd number of negative factors, the product is negative.

* **Zero as a factor :** If *any* factor is 0, the product is 0.

Exercise 1.27

Multiply:

a) $(-6)(2)(-3)(1)$

b) $(-4)(-3)(-1)(2)$

c) $256(-48)(13)(0)$

Solution 1.27

a) $(-6)(2)(-3)(1) = 36$ Two negative factors: positive product.

b) $(-4)(-3)(-1)(2) = -24$ Three negative factors: negative product.

c) $256(-48)(13)(0) = 0$ 0 as a factor makes the product 0.

MULTIPLYING AND ORDER OF OPERATIONS

The following exercise involves order of operations and multiplication of signed numbers. Remember to work inside the parentheses first, perform multiplication and division in order from left to right, and then perform addition and subtraction in order from left to right.

Exercise 1.28

Simplify:

a) $-8 + 2(-2)(-1)$

b) $(5 - 7)(6 - 10)$

c) $(-2)(-3)(-1) - 6[5 - (-3)]$

Solution 1.28

a) $-8 + 2(-2)(-1)$

$= -8 + 4$	Multiplying before adding; two negative factors: positive product.
$= -4$	Different signs: Use sign of number with larger absolute value.

b) $(5 - 7)(6 - 10)$

$= [5 + (-7)][6 + (-10)]$	Work inside the brackets first.
$= (-2)(-4)$	Different signs: Use sign of number with larger absolute value.
$= 8$	Two negative factors: positive product.

c) $(-2)(-3)(-1) - 6[5 - (-3)]$

$(-2)(-3)(-1) - 6[5 + 3]$	Work inside the brackets first.
$= (-2)(-3)(-1) - 6[8]$	
$= -6 - 6[8]$	Multiply from left to right.
$= -6 - 48$	Multiply.
$= -6 + (-48)$	Subtract.
$= -54$	Same signs: Use the common sign.

1.7 DIVIDING REAL NUMBERS

Next we extend the arithmetic concept of dividing two positive integers to dividing real numbers. We begin with the rules for dividing two signed numbers.

DIVIDING TWO SIGNED NUMBERS

There are four cases to consider when dividing two signed numbers:

1. A positive number divided by a positive number equals a positive number.
2. A negative number divided by a negative number equals a positive number.
3. A positive number divided by a negative number equals a negative number.
4. A negative number divided by a positive number equals a negative number.

Many students remember only these two rules when dividing two numbers:

* Same signs = positive quotient
* Different signs = negative quotient

Exercise 1.29

Divide:

a) $\frac{8}{2}$

b) $-6 \div -3$

c) $-5\overline{)10}$

d) $\frac{-12}{4}$

Solution 1.29

a) $\frac{8}{2} = 4$ Same signs: positive quotient.

b) $-6 \div -3 = 2$ Same signs: positive quotient.

c) $\frac{10}{-5} = -2$ Different signs: negative quotient.

d) $\frac{-12}{4} = -3$ Different signs: negative quotient.

DIVISION AND ZERO

Division is defined in terms of multiplication: a divided by b, where b does not equal to 0, is defined to be $\frac{a}{b} = a \cdot \frac{1}{b}, b \neq 0$.

Thus, each division problem can be rewritten as a multiplication problem.

Exercise 1.30

Write each division problem as a multiplication problem, and then multiply:

a) $\frac{8}{4}$

b) $-10 \div 2$

c) $\frac{0}{9}$

Solution 1.30

a) $\frac{8}{4} = 8 \cdot \frac{1}{4} = 2$

b) $-10 \div 2 = -10 \cdot \frac{1}{2} = -5$

c) $\frac{0}{9} = 0 \cdot \frac{1}{9} = 0$

Notice in Exercise (c) that 0 divided by a nonzero number equals 0 times a nonzero number, which equals 0. Therefore, 0 divided by a nonzero number equals 0

or

$\frac{0}{a} = 0$, where $a \neq 0$.

Notice in the definition of division that b, the divisor (or denominator), cannot equal 0. This leads to the following rule: Any number divided by zero is not defined

or

$\frac{a}{0}$ is not defined, for any number a.

In particular, $\frac{0}{0}$ is not defined.

Exercise 1.31

Divide, if possible:

a) $\dfrac{0}{4}$

b) $\dfrac{6}{0}$

c) $\dfrac{0-(6)}{2}$

d) $\dfrac{0}{0}$

Solution 1.31

a) $\dfrac{0}{4}=0$

b) $\dfrac{6}{0}$ is not defined. Division by 0 is not defined.

c) $\dfrac{0(-6)}{2}=\dfrac{0}{2}=0$

d) $\dfrac{0}{0}$ is not defined. Division by 0 is not defined.

DIVISION AND ORDER OF OPERATIONS

Remember that a fraction bar, which is actually a division symbol, is a grouping symbol. It separates the entire numerator from the entire denominator. Thus, you must simplify the numerator and denominator separately.

Exercise 1.32

Find each quotient:

a) $\dfrac{-10}{4-(-6)}$

b) $\dfrac{-6(-2)-8}{-10+6-(-3)}$

c) $\dfrac{6^2-3^2}{-3(4-5)}$

Solution 1.32

a) $\dfrac{-10}{4-(-6)}=\dfrac{-10}{4+6}$ — Change the subtraction symbol to +; the opposite of –6 is 6.

$=\dfrac{-10}{10}$

$=-1$ — Different signs: negative quotient.

b) $\dfrac{-6(-2)-8}{-10+6-(-3)}$

$=\dfrac{12-8}{-10+6+(3)}$ — $\dfrac{\text{Multiplying: same signs: positive product.}}{\text{Change subtraction to adding the opposite.}}$

$=\dfrac{12+(-8)}{-4+3}$ — $\dfrac{\text{Change subtraction to adding the opposite.}}{\text{Add from left to right.}}$

$=\dfrac{4}{-1}=-4$ — Different signs: negative quotient.

c) $\dfrac{6^2-3^2}{-3(4-5)}$

$=\dfrac{36-9}{-3(4+(-5))}$ — $\dfrac{\text{Exponents first.}}{\text{Work inside parentheses first.}}$

$=\dfrac{27}{-3(-1)}$ — Simplify the numerator and the denominator.

$=\dfrac{27}{3}$ — Same signs: positive product in denominator.

$=9$ — Same signs: positive quotient.

1.8 SETS OF NUMBERS

In arithmetic, students usually work with positive integers. As you continue your study of algebra, you will expand the types or sets of numbers with which you work. Some sets of numbers are described in the next section.

NUMBERS

Although the word "algebra" usually brings to mind working with letters like x and y, a strong background in working with numbers is essential to success. There are several number systems with which you should be familiar.

* The **natural numbers** (or counting numbers) are probably the first numbers you encountered in arithmetic and consist of the numbers 1, 2, 3, 4, . . . (the three dots mean to continue the pattern).

* The **whole numbers** include all of the natural numbers as well as the number 0.

* The **integers** are made up of positive integers, negative integers, and 0. Thus, the whole numbers are included in the integers.

* The **rational numbers** include all the integers, as well as the fractions you encountered in arithmetic. Remember that repeating decimals and terminating decimals can be written as fractions and are, therefore, part of the rational numbers.

* The **irrational numbers** include numbers such as π; non-repeating, non-terminating decimals; and square roots such as $\sqrt{2}$ and $\sqrt{26}$.

* The **real numbers** include both the rational numbers and irrational numbers.

The following table contains some examples of each type of number system:

Number Systems	**Examples**
Natural numbers	1, 2, 3, 4, . . .
Whole numbers	0, 1, 2, 3, . . .
Integers	–3, –2, –1, 0, 1, 2, 3, . . .
Rational numbers	0.4, –1.33, . . . , $\frac{1}{2}, -\frac{2}{5}, 1\frac{2}{3}, -4$
Irrational numbers	$\sqrt{2}, \sqrt{3}, \pi$
Real numbers	$\frac{1}{2}, \sqrt{2}, -\frac{2}{5}, 6, 3\frac{1}{3}, -3$

Exercise 1.33

List the numbers in the set

$$\left\{\sqrt{5}, 10, -\frac{1}{4}, 0, -5, 2.75, -\sqrt{7}\right\}$$

that belong to each of the following number systems:

a) Natural numbers

b) Whole numbers

c) Integers

d) Rational numbers

e) Irrational numbers

f) Real numbers

Solution 1.33

a) 10 is the only natural number in the list.

b) Whole numbers are 10 and 0.

c) Integers are 10, 0, and –5.

d) Rational numbers are $10, -\frac{1}{4}, 0, -5, 2.75$.

e) Irrational numbers are $\sqrt{5}$ and $-\sqrt{7}$.

f) Real numbers: All the numbers listed are real numbers.

Exercise 1.34

Determine whether each statement is true or false:

a) Every whole number is a rational number.

b) Every rational number is an integer.

c) A number can be both a rational number and a real number.

d) Every real number is a rational number.

e) A number can be both a whole number and a rational number.

Solution 1.34

a) True, because every whole number can be written as a fraction with a denominator of 1.

b) False. For example, $\frac{2}{3}$ cannot be written as an integer.

c) True, because the real numbers contain both the rational numbers and the irrational numbers.

d) False. For example, consider a number like $\sqrt{2}$ which is a real number but is not a rational number.

e) True. For example, consider the number 5, which can be written as $\frac{5}{1}$, which is the form of a rational number.

1.9 OPERATIONS USING RATIONAL NUMBERS

You have worked in the past with rational numbers and have called them fractions. This section blends your past work with fractions with the signed number rules and properties of real numbers. Some language you will need is reviewed in the following table:

Example	Language
$4 \cdot 5 = 20$	4 and 5 are **factors** of 20.
2, 3, 5, 7, 11, . . .	These are a few of the prime numbers. A **prime number** is a number greater than 1 whose *only* factors are 1 and the number itself.
$20 = 2 \cdot 2 \cdot 5$	20 is written here as a **product of prime factors**.
$\frac{14}{21} = \frac{2 \cdot 7}{3 \cdot 7} = \frac{2}{3}$	$\frac{14}{21}$ is written in **lowest terms** as $\frac{2}{3}$.
$\frac{2}{7}$ and $\frac{5}{7}$	Fractions with equal denominators are called **like fractions**.

REDUCING A FRACTION TO ITS LOWEST TERMS

To reduce a fraction to its lowest terms, do the following:

1. Write numerator and denominator as a product of prime factors.
2. Divide the numerator and denominator by any common factors.

Exercise 1.35

Reduce $\frac{30}{105}$ to lowest terms.

Solution 1.35

$30 = 2 \cdot 3 \cdot 5$

$105 = 3 \cdot 5 \cdot 7$

So, $\frac{30}{105} = \frac{2 \cdot 3 \cdot 5}{3 \cdot 5 \cdot 7}$ Write the numerator and denominator as a product of prime factors.

$= \frac{3 \cdot 5 \cdot 2}{3 \cdot 5 \cdot 7}$ Use the commutative and associative properties of multiplication to change the order in the numerator.

$= \frac{2}{7}$ Divide by the common factors of 3 and 5.

BASIC OPERATIONS WITH FRACTIONS

Recall the following basic rules for adding, subtracting, multiplying, and dividing fractions:

$\frac{a}{b} \cdot \frac{c}{d} = \frac{ac}{bd}$	Multiplying fractions
$\frac{a}{b} \div \frac{c}{d} = \frac{a}{b} \cdot \frac{d}{c} = \frac{ad}{bc}$	Dividing fractions
$\frac{a}{b} + \frac{c}{b} = \frac{a+c}{b}$	Adding like fractions
$\frac{a}{b} - \frac{c}{b} = \frac{a-c}{b}$	Subtracting like fractions

Exercise 1.36

Perform the indicated operations. Write your answers in lowest terms:

a) $\frac{1}{4} \cdot \frac{3}{8}$

b) $\frac{2}{5} \div \frac{4}{25}$

c) $\frac{2}{7} + \frac{3}{7}$

d) $\frac{4}{6} - \frac{1}{6}$

e) $\frac{x}{5}+\frac{3}{5}$

f) $\frac{x-2}{7}-\frac{3}{7}$

Solution 1.36

a) $\frac{1}{4}\cdot\frac{3}{8}=\frac{3}{32}$ — To multiply fractions, multiply numerators and multiply denominators.

b) $\frac{2}{5}\div\frac{4}{25}=\frac{2}{5}\cdot\frac{25}{4}$ — To divide fractions, invert the divisor, and then multiply.

$=\frac{50}{20}$ — Multiply the numerators and multiply the denominators.

$=\frac{5\cdot 10}{2\cdot 10}$ — Divide by the common factor of 10.

$=\frac{5}{2}$ — Reduce the fraction to its lowest terms.

c) $\frac{2}{7}+\frac{3}{7}=\frac{5}{7}$ — To add fractions with the same denominator, add the numerators and keep the denominator the same.

d) $\frac{4}{6}-\frac{1}{6}=\frac{3}{6}=\frac{1}{2}$ — To subtract fractions with the same denominator, subtract the numerators and keep the denominator the same. Then reduce the fraction to its lowest terms.

e) $\frac{x}{5}+\frac{3}{5}=\frac{x+3}{5}$ — To add fractions with the same denominator, add the numerators and keep the denominator the same.

f) $\frac{x-2}{7}-\frac{3}{7}=\frac{x-2-3}{7}$ — To subtract fractions with the same denominator, subtract the numerators and keep the denominator the same.

$=\frac{x-5}{7}$ — Combine similar terms.

LEAST COMMON DENOMINATORS

The rules in the preceding section for adding and subtracting fractions are for fractions with equal denominators. If the given fractions do not have equal denominators, you first have to rewrite them with a least common denominator (LCD). The least common denominator is the smallest denominator that each of the given denominators divides into evenly. While there are many common denominators, it is usually best to work with the least or smallest common denominator to keep your computations as simple as possible. There are two popular methods for finding the least common denominator.

First Method for Finding the LCD

The first method for finding the least common denominator is as follows:

1. Write a list containing multiples of each denominator. Multiples are found by multiplying by 1, 2, 3, 4, and so on.
2. Find the smallest number common to each list. This will be the least common denominator.

Exercise 1.37

Find the least common denominator for $\frac{5}{6}$ and $\frac{2}{9}$.

Solution 1.37

Multiples of 6 are 6, 12, 18, 24, 30, **36**, 42, 48, 54, 60, . . .

Multiples of 9 are 9, 18, 27, **36**, 45, 54, 63, 72, . . .

36 is the smallest number common to both lists and so is the least common denominator. Note that 54 is also common to both lists and may be used for a common denominator, but it is not the *least* common denominator.

Second Method for Finding the LCD

The second method for finding the least common denominator is as follows:

1. Write each denominator as a product of prime factors, using exponents as necessary.
2. The least common denominator is the product of each factor, raised to the highest power that appears in *any* denominator.
3. Simplify your answer.

Exercise 1.38

Find the least common denominator for $\frac{7}{8}$ and $\frac{1}{12}$.

Solution 1.38

$8 = 2 \cdot 2 \cdot 2 = 2^3$ Write each denominator as a product of prime factors.

$12 = 2 \cdot 2 \cdot 3 = 2^2 \cdot 3$ The least common denominator $= 2^3 \cdot 3$. Use each factor raised to the highest power, so use $2^3 \cdot 3$.

$2^3 \cdot 3 = 8 \cdot 3$

$= 24$

Just as fractions can be reduced by dividing the numerator and denominator by any common factors, fractions can be "built up" by multiplying the numerator and denominator by a common factor.

Exercise 1.39

Write each fraction with the indicated denominator:

a) $\frac{5}{6} = \frac{}{36}$

b) $\frac{2}{9} = \frac{}{36}$

c) $\frac{7}{8} = \frac{}{24}$

d) $\frac{1}{12} = \frac{}{24}$

Solution 1.39

a) $\frac{5}{6} = \frac{5 \cdot 6}{6 \cdot 6}$ Because $6 \cdot 6 = 36$, multiply the numerator and denominator by 6.

$= \frac{30}{36}$

b) $\frac{2}{9} = \frac{2 \cdot 4}{9 \cdot 4}$ Because $9 \cdot 4 = 36$, multiply the numerator and denominator by 4.

$= \frac{8}{36}$

c) $\frac{7}{8} = \frac{7 \cdot 3}{8 \cdot 3}$ Because $8 \cdot 3 = 24$, multiply the numerator and denominator by 3.

$= \frac{21}{24}$

d) $\frac{1}{12} = \frac{1 \cdot 2}{12 \cdot 2}$ Because $12 \cdot 2 = 24$, multiply the numerator and denominator by 2.

$= \frac{2}{24}$

ADDING AND SUBTRACTING FRACTIONS WITH DIFFERENT DENOMINATORS

You must have a common denominator to use the basic rules for adding and subtracting fractions. The following rules outline the procedure:

1. Find the least common denominator by either method listed in the preceding section.
2. Build up each fraction to have the least common denominator by multiplying the numerator and denominator by a common factor.
3. Use the basic rules for adding and subtracting fractions.

Exercise 1.40

Perform the indicated operations. Write answers in the lowest terms:

a) $\frac{5}{6} + \frac{2}{9}$

b) $\frac{7}{8} - \frac{1}{12}$

Solution 1.40

a) $\frac{5}{6} + \frac{2}{9} = \frac{5}{6} \cdot \frac{3}{3} + \frac{2}{9} \cdot \frac{2}{2}$ The least common denominator for 6 and 9 is 18. Build up each fraction to get the least common denominator.

$= \frac{15}{18} + \frac{4}{18}$ Add numerators, keep common denominator.

$= \frac{19}{18}$

b) $\frac{7}{8}-\frac{1}{12}=\frac{7}{8}\cdot\frac{3}{3}-\frac{1}{12}\cdot\frac{2}{2}$ The least common denominator of 8 and 12 is 24. Build up each fraction to get the least common denominator.

$=\frac{21}{24}-\frac{2}{24}$

$=\frac{9}{24}$ Subtract numerators; keep the common denominator.

Now that you know the rules for signed numbers, you can also use these rules with fractions.

Exercise 1.41

Perform the indicated operations. Write answers in lowest terms:

a) $\left(-\frac{8}{7}\right)\left(-\frac{2}{3}\right)$

b) $\left(-\frac{3}{8}\right)\div\left(\frac{9}{16}\right)$

c) $\left(-\frac{3}{5}\right)+\left(-\frac{3}{10}\right)$

d) $-\frac{4}{7}-\left(-\frac{2}{3}\right)$

e) $\frac{7}{8}-\frac{11}{12}$

Solution 1.41

a) $\left(-\frac{8}{7}\right)\left(-\frac{2}{3}\right)=\frac{16}{21}$ Same signs: positive product.

b) $\left(-\frac{3}{8}\right)\div\left(\frac{9}{16}\right)=\left(-\frac{3}{8}\right)\left(\frac{16}{9}\right)$ Invert the divisor and multiply.

$=-\frac{48}{72}$ Different signs: negative product.

$=-\frac{2\cdot 24}{3\cdot 24}$

$=-\frac{2}{3}$ Reduce.

c) $\left(-\frac{3}{5}\right)+\left(-\frac{3}{10}\right)=\left(-\frac{3}{5}\right)\cdot\frac{2}{2}+\left(-\frac{3}{10}\right)$ — Build up the first fraction to get the least common denominator.

$=\left(-\frac{6}{10}\right)+\left(-\frac{3}{10}\right)$ — Add numerators and keep the common denominator.

$=\frac{-6+(-3)}{10}$

$=\frac{-9}{10}$ or $-\frac{9}{10}$ — The negative symbol may be written in the numerator or in front of the fraction.

d) $-\frac{4}{7}-\left(-\frac{2}{3}\right)=-\frac{4}{7}+\frac{2}{3}$ — Subtract.

$=-\frac{4}{7}\cdot\frac{3}{3}+\frac{2}{3}\cdot\frac{7}{7}$ — Build up each fraction to get the least common denominator.

$=\frac{-12}{21}+\frac{14}{21}$

$=\frac{-12+14}{21}$ — Add numerators and keep the common denominator.

$=\frac{2}{21}$

e) $\frac{7}{8}-\frac{11}{12}=\frac{7}{8}+\left(\frac{-11}{12}\right)$ — Subtract.

$=\frac{7}{8}\cdot\frac{3}{3}+\frac{-11}{12}\cdot\frac{2}{2}$ — Build up each fraction to get the least common denominator.

$=\frac{21}{24}+\frac{-22}{24}$

$=\frac{21+(-22)}{24}$ — Add the numerators and keep the common denominator.

$=\frac{-1}{24}$ or $-\frac{1}{24}$ — The symbol may be carried in the numerator or in front of the fraction.

TEST YOURSELF

1) Simplify the following:

(a) $(3 + 7)$

(b) $[14 - 3]$

(c) $6 \cdot 2$

(d) $15 \div 5$

(e) $\frac{14 \cdot 2}{4 + 3}$

2) Find the value of the following:

(a) 3^2

(b) 1^7

(c) -4^3

(d) 2^5

(e) -2^4

3) Simplify the following:

(a) $(3 + 2) \div 5$

(b) $(2 - 3) + 6$

(c) $[(9 - 3) \div (2 + 4)]$

(d) $7(8 \div 2) + 90$

(e) $[(4 - 8) \div (6 + 3)] \div 5$

4) Simplify the following:

(a) $\{4 + [7 - (2 - 1)]\}$

(b) $(2 + 5^2 - 3 \div 3)$

(c) $5 + 9 \div 3 + 2 \div 4$

(d) $(-9)^2 \div 3 + \left(4\frac{1}{2} - \frac{3}{2}\right)$

(e) $\{6 - (2 + 4^3) + (8 \div 4^1)\}$

5) Graph the following numbers on a number line:

$3, 6, -2, -5, \frac{1}{3}, 7.9, \frac{13}{3}, -4.1$

6) Write an equivalent sentence in symbols for each of the following:

(a) The sum of x and 5 is equal to 12.

(b) The difference of y and 2 is greater than or equal to 6.

(c) The product of 3 and x is less than the sum of x and 8.

(d) The quotient of x and 2 is equal to the difference of 5 and 1.

7) Insert < or > to make each statement true:

(a) 0 ____ −2

(b) 3 ____ 6

(c) 4.6 ____ $\frac{9}{2}$

(d) $-2\frac{1}{4}$ ____ $2\frac{1}{4}$

(e) $\sqrt{9}$ ____ 3.3

8) Find the absolute value of the following:

(a) 8

(b) $-4\frac{1}{3}$

(c) 0

(d) $-\sqrt{5}$

(e) 119

9) Insert < or > to make each statement true:

(a) $|3|$ ____ $|-14|$

(b) $|8|$ ____ $|2|$

(c) $|-4|$ ____ $|7|$

(d) $|-1|$ ____ $|-9|$

(e) $|0|$ ____ $|-10|$

10) Find the opposite of each of the following:

(a) 11

(b) –11

(c) 0

(d) $-3\frac{1}{3}$

(e) 2.6

11) For each number, give the opposite and the absolute value:

(a) –6

(b) 4

(c) 0

(d) y

12) Find each sum:

(a) 6 + 3

(b) –2 + (–4)

(c) –8 + (–9)

(d) 7 + 5

(e) 1 + 14

13) Find each sum:

(a) 5 + (–3)

(b) 4 + (–8)

(c) (–7) + 10

(d) (–6) + 4

(e) 0 + (–2)

14) Find each sum:

(a) –5 + 3 + (–7)

(b) (–6 +13) + [(–1) + (–5)]

(c) $-7 + 3(-3 + 6) + 5$

(d) $\{[2 + (-5)] + [(-3) + (-4)]\}$

(e) $(10 + 11) + [(-10) + (-11)]$

15) Find each difference:

(a) $3 - 6$

(b) $-2 - 9$

(c) $14 - (-12)$

(d) $-8 - (-7)$

(e) $16 - (-16)$

16) Simplify each of the following:

(a) $1 - 5 - 4$

(b) $1 - (5 - 4)$

(c) $7 - (-2 + 4)$

(d) $(-4 - 3) - (-8 - 6)$

(e) $(-8 + 2) - (3 - 5)$

17) Simplify. Use the rules for order of operations, as necessary:

(a) $2 \cdot 5 - 4 \cdot 7$

(b) $3 \cdot 5 - 6 \cdot 8$

(c) $5 \cdot 2 - 4 \cdot 3 - 5 \cdot 4$

(d) $2 \cdot 3^2 - 4 \cdot 5^2$

(e) $4 \cdot 2^2 - 3 \cdot 4^2 - 8 \cdot 6^2$

18) Write each phrase as an equivalent expression in symbols:

(a) The difference of x and 3 is equal to 10.

(b) The product of 5 and y is greater than the difference of 4 and 7.

(c) The sum of t and 5 is less than the difference of $2t$ and 3.

(d) 4 subtracted from x is equal to 8.

(e) The sum of x and 5 is greater than or equal to the difference of 10 and x.

19) State the property (or properties) that justifies each statement:

(a) $-3 + 4 = 4 + (-3)$

(b) $7 + 0 = 7$

(c) $18(3 + x) = 18(3) + 18(x)$

(d) $296(1) = 296$

(e) $5(72) = 72(5)$

20) Apply the distributive property to rewrite each of the following. Simplify, when possible:

(a) $3(7x + 9)$

(b) $6(a + b)$

(c) $-4(2x + 3y)$

(d) $-(12s + 15t)$

(e) $8(3x - 2)$

(f) $5(3x - 2)$

(g) $2(7y + 4) - 8$

(h) $5(3x - 4) + 6$

21) Use the associative property to rewrite each expression, and then simplify:

(a) $5 + (3 + x)$

(b) $3(2x)$

(c) $(y + 4) + 2$

(d) $5(7y)$

22) State the property (or properties) that justifies each statement:

(a) $3 + (7 + 4) = 7 + (3 + 4)$

(b) $b(2a) = 2(ab)$

(c) $(2 + 3) + 5 = 5\ (3 + 2)$

(d) $5(4x) = (4 \cdot 5)x$

23) Find each product:

(a) $3(11)$

(b) $(-7)(2)$

(c) $(-6)(-4)$

(d) $5(-3)$

(e) $(8)(1)$

24) Find each product:

(a) $(13)(12)(11)(10)$

(b) $(5)(-4)(0)(-9)$

(c) $(-6)(8)(-1)(-2)$

(d) $(-2)(-3)(-8)(0)$

(e) $(-1)(-2)(-1)(-2)$

25) Simplify:

(a) $6 + (-3)(-4)$

(b) $(7 + 3)(7 - 3)$

(c) $(2)(0)(3) -4[(2)(-1) + 7]$

(d) $\{6[4 + (-3)] -2(4)\}$

(e) $4 - 2 - (-1 - 6)$

26) Divide:

(a) $\frac{9}{3}$

(b) $(-4) \div (-2)$

(c) $36 \div 6$

(d) $-\frac{15}{5}$

(e) $(-7) \div (-1)$

27) Rewrite each division problem as a multiplication problem and simplify:

(a) $\frac{9}{3}$

(b) $(-4) \div (-2)$

(c) $36 \div 6$

(d) $-\frac{15}{5}$

(e) $(-7) \div (-1)$

28) Divide, if possible:

(a) $\frac{0}{5}$

(b) $\frac{3}{0}$

(c) $\frac{0}{0}$

(d) $\frac{0(-2)}{4}$

(e) $\frac{4-4}{3}$

29) Simplify:

(a) $\frac{2^5}{-8(1+3)}$

(b) $\frac{6+2^2}{3-(-7)}$

(c) $\frac{2(0)+5}{-14+7-(-2)}$

(d) $\frac{-14}{3+5+(-1)}$

(e) $\frac{6^2-3^2}{[3(-4)+9]}$

(f) $\frac{2(-4)+5(-3)}{8-5}$

(g) $\dfrac{-2\cdot 5^2-4\cdot 3^2}{-4(-2)}$

(h) $\dfrac{|2-7|}{2-7}$

30) List the numbers in the set $\{-3.21, \sqrt{3}, \frac{1}{5}, 0, 6, -\sqrt{11}, -7\}$ that belong to each of the following:

(a) Irrational numbers

(b) Real numbers

(c) Natural numbers

(d) Rational numbers

(e) Integers

(f) Whole numbers

31) Determine whether each statement is true or false:

(a) Zero is a whole number and a rational number.

(b) Some numbers are both rational numbers and irrational numbers.

(c) Every whole number is an integer.

(d) A whole number can be a negative integer.

32) Perform the indicated operations. Write your answers in lowest terms:

(a) $\frac{4}{3}\cdot\frac{2}{7}$

(b) $\frac{1}{3}\div\frac{4}{9}$

(c) $\frac{4}{5}+\frac{7}{5}$

(d) $\frac{8}{7}-\frac{6}{7}$

(e) $\frac{0}{6}+\frac{7}{6}$

33) Find the least common denominator:

(a) $\frac{3}{6}$ and $\frac{7}{9}$

(b) $\frac{2}{5}$ and $\frac{1}{15}$

(c) $\frac{7}{4}$ and $\frac{3}{7}$

(d) $\frac{5}{2}$ and $\frac{4}{3}$

(e) $\frac{1}{11}$ and $\frac{6}{13}$

34) Write each fraction with indicated denominator:

(a) $\frac{5}{7} = \frac{}{49}$

(b) $\frac{3}{8} = \frac{}{24}$

(c) $\frac{2}{9} = \frac{}{36}$

(d) $\frac{1}{3} = \frac{}{12}$

(e) $\frac{2}{3} = \frac{}{21}$

35) Perform the indicated operations:

(a) $\frac{3}{4} - \frac{1}{3}$

(b) $\frac{2}{7} + \frac{3}{14}$

(c) $\frac{8}{3} + \frac{7}{2}$

(d) $\frac{2}{9} + \frac{5}{6}$

(e) $\frac{8}{5} - \frac{2}{3}$

36) Perform the indicated operations:

(a) $\left(-\frac{7}{8}\right)\left(-\frac{2}{3}\right)$

(b) $\left(-\frac{5}{6}\right) \div \left(\frac{15}{2}\right)$

(c) $\left(\frac{1}{6}\right) + \left(\frac{-7}{2}\right)$

(d) $\frac{2}{-3} - \left(\frac{4}{-7}\right)$

(e) $\frac{7}{9} - \frac{5}{6}$

TEST YOURSELF ANSWERS

1) (a) 10

(b) 11

(c) 12

(d) 3

(e) 4

2) (a) 9

(b) 1

(c) –64

(d) 32

(e) –16

3) (a) 1

(b) 5

(c) 1

(d) 118

(e) $-\frac{4}{45}$

4) (a) 10

(b) 26

(c) $\frac{17}{2}$

(d) 30

(e) –58

5) 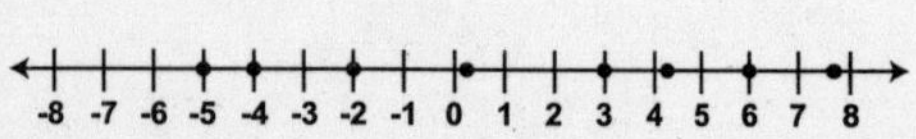

6) (a) $x + 5 = 12$

(b) $y - 2 \geq 6$

(c) $3x < x + 8$

(d) $\frac{x}{2} = 5 - 1$

7) (a) >

(b) <

(c) >

(d) <

(e) <

8) (a) 8

(b) $4\frac{1}{3}$

(c) 0

(d) $\sqrt{5}$

(e) 119

9) (a) <

(b) >

(c) <

(d) <

(e) <

10) (a) -11

(b) 11

(c) 0

(d) $3\frac{1}{3}$

(e) -2.6

11) (a) $6, 6$

(b) $-4, 4$

(c) $0, 0$

(d) $-y, |y|$

12) (a) 9

(b) -6

(c) -17

(d) 12

(e) 15

13) (a) 2

(b) -4

(c) 3

(d) -2

(e) -2

14) (a) -9

(b) 1

(c) 7

(d) -10

(e) 0

15) (a) -3

(b) -11

(c) 26

(d) -1

(e) 32

16) (a) -8

(b) 0

(c) 5

(d) 7

(e) -4

17) (a) -18

(b) -33

(c) -22

(d) -82

(e) -320

18) (a) $x - 3 = 10$

(b) $5y > 4 - 7$

(c) $t + 5 < 2t - 3$

(d) $x - 4 = 8$

(e) $x + 5 \geq 10 - x$

19) (a) Commutative property

(b) Additive identity property

(c) Distributive property

(d) Multiplicative identity property

(e) Commutative property

20) (a) $3(7x) + 3(9) = 21x + 27$

(b) $6(a) + 6(b) = 6a + 6b$

(c) $(-4)(2x) + (-4)(3y)$

$= -8x - 12y$

(d) $(-1)12s + (-1)15t = -12s - 15t$

(e) $8(3x) - 8(2) = 24x - 16$

(f) $5(3x) - 5(2) = 15x - 10$

(g) $2(7y) + 2(4) - 8 = 14y$

(h) $5(3x) - 5(4) + 6 = 15x - 14$

21) (a) $(5 + 3) + x = 8 + x$

(b) $(3 \cdot 2)x = 6x$

(c) $y + (4 + 2) = y + 6$

(d) $(5 \cdot 7)y = 35y$

22) (a) Commutative and associative

(b) Commutative and associative

(c) Commutative and commutative

(d) Associative and commutative

23) (a) 33

(b) –14

(c) 24

(d) –15

(e) 8

24) (a) 17,160

(b) 0

(c) –96

(d) 0

(e) 4

25) (a) 18

(b) 40

(c) –20

(d) –2

(e) 9

26) (a) 3

(b) 2

(c) 6

(d) –3

(e) 7

27) (a) $9 \cdot \frac{1}{3} = 3$

(b) $-4 \cdot -\frac{1}{2} = 2$

(c) $36 \cdot \frac{1}{6} = 6$

(d) $-15 \cdot \frac{1}{5} = -3$

(e) $-7 \cdot (-1) = 7$

28) (a) 0

(b) Undefined

(c) Undefined

(d) 0

(e) 0

29) (a) -1

(b) 1

(c) -1

(d) -2

(e) -9

(f) $-\frac{23}{3}$

(g) $-\frac{43}{4}$

(h) -1

30) (a) $\sqrt{3}, -\sqrt{11}$

(b) $-3.21, \sqrt{3}, \frac{1}{5}, 0, 6, -\sqrt{11}, -7$

(c) 6

(d) $-3.21, \frac{1}{5}, 0, 6, -7$

(e) $0, 6$

(f) $0, 6$

31) (a) True

(b) False

(c) True

(d) False

32) (a) $\frac{8}{21}$

(b) $\frac{3}{4}$

(c) $\frac{11}{5}$

(d) $\frac{2}{7}$

(e) $\frac{7}{6}$

33) (a) 18

(b) 15

(c) 28

(d) 6

(e) 143

34) (a) $\frac{35}{49}$

(b) $\frac{9}{24}$

(c) $\frac{8}{36}$

(d) $\frac{4}{12}$

(e) $\frac{14}{21}$

35) (a) $\frac{5}{12}$

(b) $\frac{1}{2}$

(c) $\frac{37}{6}$

(d) $\frac{19}{18}$

(e) $\frac{14}{15}$

36) (a) $\frac{7}{12}$

(b) $-\frac{1}{9}$

(c) $-\frac{10}{3}$

(d) $-\frac{2}{21}$

(e) $-\frac{1}{18}$

CHAPTER 2

Solving Linear Equations and Inequalities

The information in many basic word problems can be expressed using linear equations and inequalities. This chapter introduces the skills necessary to solve linear equations and inequalities.

2.1 SIMPLIFYING AND EVALUATING EXPRESSIONS

Before beginning our discussion about simplifying and evaluating expressions, you need to be familiar with the following vocabulary:

Word	Meaning	Examples
Constant	A number	5, –2
Variable	A letter used to represent an unknown number	x, y, z, a, b, c
Term	A number or product of number(s) and variable(s) raised to powers	$4, 2xy, -x^2, 5s^2r$
Numerical coefficient	The number part of a term	In $2xy$, 2 is the numerical coefficient.
Similar (or like) terms	Terms with the same variable(s) and exponent(s)	$-6xy$ and $10xy$ $4x$ and $-3x$ x^2y and $5x^2y$

Note: The numerical coefficient of 1 does not need to be written. Thus, x means $1 \cdot x$, a means $1 \cdot a$, $-x$ means $-1 \cdot x$, $-a$ means $-1 \cdot a$.

ADDING AND SUBTRACTING SIMILAR TERMS

To combine similar terms:

1. Add or subtract the numerical coefficients.
2. Write the same variable part.

Remember that only similar terms can be added or subtracted.

Exercise 2.1

Simplify by combining similar terms:

a) $-7xy - 9xy$

b) $x^2 - 5x^2$

c) $-4a + 7a + 2 - 5$

d) $2p^2 - 3p + 1 + 6p^2 - 8 + 4p$

e) $6x + 4x^2$

Solution 2.1

a) $-7xy - 9xy = (-7 - 9)xy$	Combine numerical coefficients.
$= -16xy$	Write the same variable part.
b) $x^2 - 5x^2 = 1x^2 - 5x^2$	x^2 means $1x^2$.
$= (1 - 5)x^2$	Combine numerical coefficients.
$= -4x^2$	Write the same variable part.
c) $-4a + 7a + 2 - 5$	
$= (-4 + 7)a + 2 - 5$	Combine numerical coefficients of similar terms.
$= 3a + 2 - 5$	Write the same variable part.
$= 3a - 3$	Combine constants.
d) $2p^2 - 3p + 1 + 6p^2 - 8 + 4p$	
$= (2 + 6)p^2 + (-3 + 4)p + 1 - 8$	Combine numerical coefficients of similar terms.
$= 8p^2 + 1p - 7$	Write same variable part, combine constants.
$= 8p^2 + p - 7$	$1p$ is the same as p.
e) $6x + 4x^2$	This expression cannot be simplified any further, because the two terms are not similar and therefore, cannot be added.

SIMPLIFYING EXPRESSIONS

An **algebraic expression** contains numbers, variables, and one or more of the four basic operations (addition, subtraction, multiplication, division). For example, $5xy + 8x$, $5(2x + y)$, and $x^2 - 3x + 2$ are algebraic expressions. To simplify expressions, we reduce the number of terms in the expression.

To simplify an expression, do the following:

1. Follow order of operations (simplify inside grouping symbols, simplify exponents, multiply and divide in order from left to right, and then add and subtract in order from left to right).
2. Remove the parentheses, if necessary, using the distributive property.
3. Combine similar terms.

Exercise 2.2

Simplify:

a) $4(2x - 8)$
b) $-2(3m + 5)$
c) $-3(y - 2) + 4(2y + 5)$
d) $6 - 5(4a + 3)$

Solution 2.2

a) $4(2x - 8) = 4(2x) - 4(8)$ — Use the distributive property to remove the parentheses.

$= 8x - 32$ — Multiply. Note that $8x$ and 32 are not similar terms and, therefore, cannot be combined any further.

b) $-2(3m + 5) = -2(3m) + (-2)(5)$ — Use the distributive property.

$= -6m + (-10)$ — Multiply.

$= -6m - 10$ — We usually write $+ (-10)$ as -10.

c) $-3(y - 2) + 4(2y + 5)$

$= -3(y) - 3(-2) + 4(2y) + 4(5)$ — Use the distributive property.

$= -3y + 6 + 8y + 20$ — Multiply. Note that $(-3)(-2) = +6$, because a negative times a negative is a positive.

$= -3y + 8y + 6 + 20$ — Use the commutative property to change order.

$= (-3 + 8)y + (6 + 20)$ — Combine similar terms and combine constants.

$= 5y + 26$

d) $6 - 5(4a + 3) = 6 + (-5)(4a) + (-5)(3)$ — Follow order of operations. You *must* use the distributive property to multiply *before* subtracting $6 - 5$.

$= 6 + (-20a) + (-15)$ — Multiply.

$= -20a + 6 + (-15)$ — Use the commutative property to change the order of 6 and $-20a$.

$= -20a - 9$ — Combine constants. Stop here because $-20a$ and 9 are not similar terms.

EVALUATING EXPRESSIONS

We stated earlier that variables are used to represent unknown numbers. Once we know the value of a variable, we can replace that variable with its value.

***Remember*:** Always replace a variable with a value by putting the value in parentheses.

Exercise 2.3

Evaluate each expression using the given value of the variable(s):

a) $2x - 5$; $x = 3$
b) $4 - 6y$; $y = -2$
c) $(2p + 3q)(-2m)$; $p = -1$, $q = 2$, $m = -3$
d) $x^2 - 3x + 2$; $x = 4$

Solution 2.3

a) $2x - 5 = 2(3) - 5$ — Replace the variable with the number 3, using parentheses.
$= 6 - 5$ — Multiply before adding or subtracting.
$= 1$

b) $4 - 6y = 4 - 6(-2)$ — Replace the variable with the number −2, using parentheses.
$= 4 - (-12)$ — Multiply before adding or subtracting.
$= 4 + 12$
$= 16$

c) $(2p + 3q)(-2m) = [2(-1) + 3(2)][-2(-3)]$ — Replace each variable with the corresponding number, using parentheses.
$= [-2 + 6][6]$ — Work inside brackets first.
$= [4][6] = 24$ — No symbol between brackets implies multiplication.

d) $x^2 - 3x + 2 = (4)^2 - 3(4) + 2$ — Replace the variable with the number 4, using parentheses.
$= 16 - 3(4) + 2$ — Simplify exponents first.
$= 16 - 12 + 2$ — Multiply before adding or subtracting.
$= 4 + 2$ — Add and subtract in order from left to right.
$= 6$

Exercise 2.4

Translate each phrase into an algebraic expression using x as the unknown number. Then find the value of the expression when $x = -3$:

a) The sum of a number and 6 is decreased by 4.
b) The difference of 7 and a number is increased by 8.
c) The product of twice a number and 5 is then subtracted from 2.
d) Four more than a number is added to 12.
e) One-third of a number is decreased by 6.

Solution 2.4

a) "The sum of a number and 6" translates to $(x + 6)$. "Decreased by 4" means to subtract 4 from it:
$(x + 6) - 4$. Now, use $x = -3$:
$((-3) + 6) - 4 = (3) - 4 = -1$

b) "The difference of 7 and a number" means $(7 - x)$. "Increased by 8" means 8 is added to it:
$(7 - x) + 8$. Now, use $x = -3$:
$(7 - (-3)) + 8 = (7 + 3) + 8 = 10 + 8 = 18$

c) "The product of twice a number and 5" means $(2x \cdot 5)$. "Subtracted from 2" means the 2 is written first:
$2 - (2x \cdot 5) = 2 - 10x$. Now, use $x = -3$:
$2 - 10(-3) = 2 + 30 = 32$

d) "Four more than a number" means $(4 + x)$. "Added to 12" means $(4 + x) + 12$. Now, use $x = -3$:
$(4 + (-3)) + 12 = (1) + 12 = 13$

e) "One-third of a number" means $\frac{1}{3}x$. "Decreased by 6" means $\frac{1}{3}x - 6$. Now, use $x = -3$:

$$\frac{1}{3}(-3) - 6 = -1 - 6 = -7$$

2.2 ADDITION PROPERTY OF EQUALITY

An **equation** contains the following components:

Equality symbol
↓
$\boxed{2x+6} = \boxed{x-4}$
↓ ↓
Left side Right side

A **solution** to an equation is a number that replaces the variable and makes both sides equal. **Equivalent equations** are equations with the same solution(s). We use the **addition property of equality** to solve equations; you may add the same quantity to both sides of an equation to produce equivalent equations.

Exercise 2.5

Solve and check each equation:

a) $x - 4 = 7$

b) $y - 8 = -6$

c) $m - 1 = -3$

d) $x - \frac{2}{3} = \frac{1}{4}$

Solution 2.5

a) $x - 4 + \mathbf{4} = 7 + \mathbf{4}$ Add 4 to both sides of the equation.
$x + 0 = 11$
$x = 11$ 11 is the proposed solution to the given equation.

Check: We check a proposed solution to be sure it makes both sides of the equation equal. Later, we check to make sure the solution makes sense.

$x - 4 = 7$ This is the original equation.
$(11) - 4 \stackrel{?}{=} 7$ Replace the variable with its proposed solution. The question mark means we are checking.
$7 = 7$ Both sides are equal, so the solution checks.

Because 11 makes both sides of the equation equal, $x = 11$ is the solution to the equation.

b) $y - 8 + \mathbf{8} = -6 + \mathbf{8}$ Add 8 to both sides of the equation.
$y + 0 = 2$
$y = 2$ 2 is the proposed solution to the given equation.

Check:

$y - 8 = -6$ This is the original equation.
$(2) - 8 \stackrel{?}{=} -6$ Replace the variable with its proposed solution.
$-6 = -6$ Both sides are equal, so $y = 2$ is the solution.

c) $m - 1 + \mathbf{1} = -3 + \mathbf{1}$ Add 1 to both sides of the equation.
$m + 0 = -2$
$m = -2$ −2 is the proposed solution to the given equation.

Check:

$m - 1 = -3$	This is the original equation.
$(-2) - 1 \stackrel{?}{=} -3$	Replace the variable with its proposed solution.
$-3 = -3$	Both sides are equal, so $m = -2$ is the solution.

d)

$x - \frac{2}{3} = \frac{1}{4}$	This is the original equation.
$x - \frac{2}{3} + \mathbf{\frac{2}{3}} = \frac{1}{4} + \mathbf{\frac{2}{3}}$	Add $\frac{2}{3}$ to both sides of the equation.
$x = \frac{1}{4} \cdot \frac{3}{3} + \frac{2}{3} \cdot \frac{4}{4}$	To add fractions, write each over the LCD, 12.
$x = \frac{3}{12} + \frac{8}{12}$	Multiply each fraction.
$x = \frac{11}{12}$	Add numerators and write them over the LCD.

Check:

$x - \frac{2}{3} = \frac{1}{4}$	This is the original equation.
$\frac{11}{12} - \frac{2}{3} = \frac{1}{4}$	Replace x with $\frac{11}{12}$.
$\frac{11}{12} - \frac{2}{3} \cdot \frac{4}{4} = \frac{1}{4}$	Get a common denominator to subtract fractions.
$\frac{11}{12} - \frac{8}{12} = \frac{1}{4}$	
$\frac{1}{4} = \frac{1}{4}$	Both sides are equal, so $x = \frac{11}{12}$ is the solution.

Your goal in solving these equations is to get the variable alone on one side of the equation using the addition property of equality. Because subtraction is defined as adding an inverse, this property also can be used to subtract the same quantity from both sides of an equation.

Exercise 2.6

Solve and check each of the following:

a) $x + 5 = 8$
b) $7y = 6y - 3$
c) $2a - 4 = -6 + a$
d) $4k + 2k - 3 = -8k + 6 + 13k$
e) $3(2m - 1) - 5m = 4$
f) $4x - 2.5 = 3x + 6.2$

Solution 2.6

a) $x + 5 = 8$

$x + 5 - \mathbf{5} = 8 - \mathbf{5}$	Subtract 5 from both sides of the equation.
$x + 0 = 3$	
$x = 3$	*x* is alone on one side of the equation.
Check:	
$x + 5 = 8$	This is the original equation.
$(3) + 5 \stackrel{?}{=} 8$	Replace the variable with its proposed solution.
$8 = 8$	Both sides are equal, so the solution checks.
$x = 3$	3 is the solution to the original equation.

b) $7y = 6y - 3$

$7y - \mathbf{6y} = 6y - 3 - \mathbf{6y}$	To get the variable on one side, subtract 6*y* from both sides of the equation.
$1y = -3$	
$y = -3$	Remember that 1*y* is the same as *y*.
Check:	
$7y = 6y - 3$	This is the original equation.
$7(-3) \stackrel{?}{=} 6(-3) - 3$	Replace the variable with its proposed solution.
$-21 \stackrel{?}{=} -18 - 3$	
$-21 = -21$	Both sides are equal, so $y = -3$ is the solution.

c) $2a - 4 = -6 + a$

$2a - 4 - \boldsymbol{a} = -6 + a - \boldsymbol{a}$	To get the variable on one side, subtract *a* from both sides of the equation.
$a - 4 = -6$	
$a - 4 + \mathbf{4} = -6 + \mathbf{4}$	To get constants on one side, add 4 to both sides of the equation.
$a = -2$	

Check:

$2a - 4 = -6 + a$	This is the original equation.
$2(-2) - 4 \stackrel{?}{=} -6 + (-2)$	Replace the variable with its proposed solution.
$-4 - 4 \stackrel{?}{=} -8$	
$-8 = -8$	Both sides are equal, so $a = -2$ is the solution.

d) $4k + 2k - 3 = -8k + 6 + 13k$

$6k - 3 = 5k + 6$	Begin by adding similar terms on each side.
$6k - 3 - \mathbf{5k} = 5k + 6 - \mathbf{5k}$	To get the variable on one side, subtract $5k$ from both sides of the equation.
$k - 3 = 6$	Combine similar terms.
$k - 3 + \mathbf{3} = 6 + \mathbf{3}$	To get constants on one side, add 3 to both sides of the equation.
$k = 9$	

Check: Try this check on your own. $k = 9$ is the solution.

e) $3(2m - 1) - 5m = 4$

$6m - 3 - 5m = 4$	Use the distributive property.
$m - 3 = 4$	Subtract similar terms on the left side.
$m - 3 + \mathbf{3} = 4 + \mathbf{3}$	To get constants on one side, add 3 to both sides of the equation.
$m = 7$	

Check: Try this check on your own. $m = 7$ is the solution.

f) $4x - 2.5 = 3x + 6.2$ This is the original equation.

$4x - 2.5 - \mathbf{3x} = 3x + 6.2 - \mathbf{3x}$	To get the variable alone on one side, subtract $3x$ from both sides of the equation.
$x - 2.5 = 6.2$	Simplify each side.
$x - 2.5 + \mathbf{2.5} = 6.2 + \mathbf{2.5}$	To get constants on one side, add 2.5 to both sides.
$x = 8.7$	Add.

Check:

$4x - 2.5 = 3x + 6.2$	This is the original equation.
$4(8.7) - 2.5 = 3(8.7) + 6.2$	Replace x with 8.7.
$34.8 - 2.5 = 26.1 + 6.2$	Multiply.
$32.3 = 32.3$	Simplify each side.
	Both sides are equal, so the solution checks.

2.3 MULTIPLICATION PROPERTY OF EQUALITY

We used the additive inverse in applying the addition property of equality. We will need the reciprocal for the next step in solving equations.

Recall that the **reciprocal** or **multiplicative inverse** of a number a is $\frac{1}{a}$, $a \neq 0$.

Number	Reciprocal or Multiplicative Inverse
4	$\frac{1}{4}$
$\frac{1}{2}$	$\frac{2}{1}=2$
-3	$-\frac{1}{3}$
$-\frac{2}{3}$	$-\frac{3}{2}$

An equation like $3x = 15$ cannot be solved with the addition property of equality, but we have the multiplication property of equality to help us; you may multiply both sides of an equation by the same nonzero quantity to produce equivalent equations.

Because division is defined in terms of multiplication, this property can also be used to divide both sides of an equation by the same quantity. Note that division by 0 is undefined. Therefore, you cannot divide both sides by 0.

Exercise 2.7

Solve $\frac{1}{3}x = 6$.

Solution 2.7

$3 \cdot \frac{1}{3}x = 3 \cdot 6$	Multiply both sides of the equation by 3.
$1x = 18$	Multiply.
$x = 18$	$1x$ is the same as x.

You are probably asking, how did we decide to multiply by 3?

Remember: Multiply by the reciprocal of the numerical coefficient of the variable.

Exercise 2.8

Solve and check:

a) $3x = 15$

b) $-4m = 8$

c) $\frac{a}{3} = 6$

d) $-\frac{2}{3}y = 12$

Solution 2.8

a) $3x = 15$

$\frac{1}{3} \cdot 3x = \frac{1}{3} \cdot 15$	The reciprocal of 3 is $\frac{1}{3}$. Multiply both sides by $\frac{1}{3}$.
$x = 5$	This is the proposed solution.

or

$3x = 15$	
$\frac{3x}{3} = \frac{15}{3}$	Divide both sides by 3.
$x = 5$	This is the proposed solution.

Check:

$3x = 15$	This is the original equation.
$3(5) \stackrel{?}{=} 15$	Replace the variable with its proposed solution.
$15 = 15$	Both sides are equal, so $x = 5$ is the solution.

It may be easier in some cases to divide by the coefficient of the variable, and in other cases to multiply by the reciprocal of the coefficient.

b) $-4m = 8$

$-\frac{1}{4} \cdot -4m = -\frac{1}{4} \cdot 8$	The reciprocal of -4 is $-\frac{1}{4}$.
$m = -\frac{8}{4}$	A negative times a negative is a positive.
$m = -2$	

or

$-4m = 8$

$\dfrac{-4m}{-4} = \dfrac{8}{-4}$ Divide both sides by –4.

$m = -2$

Check:

$-4m = 8$ This is the original equation.

$-4(-2) \stackrel{?}{=} 8$ Replace the variable with its proposed solution.

$8 = 8$ Both sides are equal, so $m = -2$ is the solution.

c) $\dfrac{a}{3} = 6$

$\dfrac{1}{3}a = 6$ First rewrite $\dfrac{a}{3}$ as $\dfrac{1}{3}a$.

$3 \cdot \dfrac{1}{3}a = 3 \cdot 6$ Multiply both sides of the equation by 3.

$a = 18$

Check:

$\dfrac{a}{3} = 6$ This is the original equation.

$\dfrac{(18)}{3} \stackrel{?}{=} 6$ Replace the variable with its proposed solution.

$6 = 6$ Both the sides are equal, so the solution checks.

$a = 18$ This is the solution to the original equation.

d) $-\dfrac{2}{3}y = 12$

$-\dfrac{3}{2} \cdot -\dfrac{2}{3}y = -\dfrac{3}{2} \cdot 12$ The reciprocal of $-\dfrac{2}{3}$ is $-\dfrac{3}{2}$.

$1y = -18$ $-\dfrac{3}{2} \cdot 12 = -\dfrac{3}{2} \cdot \dfrac{12}{1} = -\dfrac{3 \cdot 12}{2} = -\dfrac{36}{2} = -18$

$y = -18$

Check:

$-\dfrac{2}{3}y = 12$ This is the original equation.

$-\dfrac{2}{3}(-18) \stackrel{?}{=} 12$ Replace the variable with its proposed solution.

$-\frac{3}{2}\cdot-\frac{18}{1}\overset{?}{=}12$ $\quad -18=-\frac{18}{1}$

$-\frac{2}{3}\cdot-\frac{18}{1}\overset{?}{=}12$ $\quad$ Multiply numerators, multiply denominators.

$\frac{36}{3}\overset{?}{=}12$ $\quad$ Reduce $\frac{36}{3}$.

$12=12$ $\quad$ Both sides are equal, so the solution checks.

$y=-18$ $\quad$ This is the solution to the original equation.

If the equation has similar terms that can be combined on one side of the equation, that step should be done before multiplying by the reciprocal. Study the following exercise:

Exercise 2.9

Solve each equation by simplifying each side first:

a) $2x-5x+7x=8$

b) $-5-7=4y-2y+y$

c) $\frac{1}{2}a+\frac{1}{4}a-\frac{3}{8}a=-6$

Solution 2.9

a) $2x-5x+7x=8$ $\quad$ This is the original equation.

$4x=8$ $\quad$ Combine similar terms.

$\frac{4x}{4}=\frac{8}{4}$ $\quad$ Divide both sides by 4.

$1x=2$ $\quad$ Reduce each fraction.

$x=2$ $\quad$ $1x=x$.

b) $-5-7=4y-2y+y$ $\quad$ This is the original equation.

$-12=3y$ $\quad$ Combine similar terms on the left side; combine similar terms on the right side.

$\frac{1}{3}\cdot-12=\frac{1}{3}\cdot3y$ $\quad$ Multiply both sides by the reciprocal of 3, $\frac{1}{3}$.

$-4=y$ or $y=-4$ $\quad$ $\frac{1}{3}\cdot-12=\frac{1}{3}\cdot\frac{-12}{1}=\frac{-12}{3}=-4$.

c) $\frac{1}{2}a+\frac{1}{4}a-\frac{3}{8}a=-6$	This is the original equation.
$\frac{4}{4}\cdot\frac{1}{2}a+\frac{2}{2}\cdot\frac{1}{4}a-\frac{3}{8}a=-6$	Multiply each fraction to get the common denominator, 8.
$\frac{4}{8}a+\frac{2}{8}a-\frac{3}{8}a=-6$	Write each fraction with the common denominator.
$\frac{3}{8}a=-6$	Combine similar terms.
$\frac{8}{3}\cdot\frac{3}{8}a=\frac{8}{3}\cdot-6$	Multiply both sides by $\frac{8}{3}$, which is the reciprocal of $\frac{3}{8}$.
$a=-16$	This is the solution.

2.4 SOLVING LINEAR EQUATIONS

Let us now use the various properties together to solve linear equations. A **linear equation** can be written in the form $ax + b = c$, where a, b, and c are real numbers. The following table gives examples of linear equations and of equations that are not linear:

Linear Equations	Nonlinear Equations
$4x + 6 = 2$	$x^2 + 2x + 6 = 0$. Linear equations have variables with exponents of 1 only.
$2x + 3 = 5x - 4$	$\sqrt{2x-1}=4$. Linear equations do not have variables under square root symbols.
$\frac{x}{4}=8$	$\frac{1}{x+3}+1=6$. Linear equations do not have variables in the denominator.
$2(x - 3) + 6 = -(x + 1) - 5$	$x^3 - 8 = 0$. Linear equations have variables with exponents of 1 only.

To solve linear equations, do the following:

1. Simplify the left side of the equation by removing parentheses, and adding and subtracting similar terms.
2. Simplify the right side of the equation by removing parentheses, and adding and subtracting similar terms.

3. Clear fractions by multiplying both sides of the equation by the LCD. Clear decimals by multiplying both sides of the equation by the appropriate power of 10.

4. Use the addition property of equality to gather the variable terms together on one side of the equation and the numbers on the other side of the equation.

5. Multiply both sides by the reciprocal of the numerical coefficient of the variable (or divide both sides by the numerical coefficient of the variable).

Exercise 2.10

Solve and check each of the following:

a) $4x - 3 - 2x = 4 - x + 5$
b) $-2(x + 3) = 6(4 - 2x)$
c) $-2(4s - 6) + 8 = -3(s + 2) + 6$
d) $3(6 - 2n) - 2(3n + 4) = 16$
e) $4(p - 2) + 3p = -6 + 3p - 2$

Solution 2.10

a) $4x - 3 - 2x = 4 - x + 5$

$2x - 3 = -x + 9$	Combine similar terms.
$2x - 3 + \mathbf{x} = -x + 9 + \mathbf{x}$	Add x to both sides.
$3x - 3 = 9$	Combine similar terms.
$3x - 3 + \mathbf{3} = 9 + \mathbf{3}$	Add 3 to both sides.
$3x = 12$	Combine similar terms.
$\frac{3x}{3} = \frac{12}{3}$	Divide both sides by the numerical coefficient.
$x = 4$	This is the proposed solution.
Check:	
$4x - 3 - 2x = 4 - x + 5$	This is the original equation.
$4(4) - 3 - 2(4) \stackrel{?}{=} 4 - (4) + 5$	Replace the variable with its proposed solution.
$16 - 3 - 8 \stackrel{?}{=} 4 - 4 + 5$	Simplify each side. Be careful to follow the order of operations.
$5 = 5$	Both sides are equal, so $x = 4$ is the solution.

b) $-2(x + 3) = 6(4 - 2x)$

$-2x - 6 = 24 - 12x$	Use the distributive property to remove the parentheses.
$-2x - 6 + \mathbf{12x} = 24 - 12x + \mathbf{12x}$	Add $12x$ to both sides.
$10x - 6 = 24$	Combine similar terms.

$10x - 6 + \mathbf{6} = 24 + \mathbf{6}$ Add 6 to both sides.

$10x = 30$

$\frac{10x}{10} = \frac{30}{10}$ Divide both sides by 10.

$x = 3$ This is the proposed solution.

Check: Try the check on your own. $x = 3$ is the solution to the original equation.

c) $-2(4s - 6) + 8 = -3(s + 2) + 6$

$-8s + 12 + 8 = -3s - 6 + 6$ Use the distributive property to remove the parentheses.

$-8s + 20 = -3s$ Simplify the left side; simplify the right side by combining similar terms.

$-8s + 20 + \mathbf{3s} = -3s + \mathbf{3s}$ Add 3*s* to both sides of the equation.

$-5s + 20 = 0$ Combine similar terms.

$-5s + 20 - \mathbf{20} = 0 - \mathbf{20}$ Subtract 20 from both sides of the equation.

$-5s = -20$ Combine similar terms.

$\frac{-5s}{-5} = \frac{-20}{-5}$ Divide both sides by –5.

$s = 4$

Check: Try the check on your own. $s = 4$ is the solution to the original equation.

d) $3(6 - 2n) - 2(3n + 4) = 16$

$18 - 6n - 6n - 8 = 16$ Use the distributive property to remove the parentheses.

$-12n + 10 = 16$ Simplify the left side by combining similar terms.

$-12n + 10 - \mathbf{10} = 16 - \mathbf{10}$ Subtract 10 from both sides.

$-12n = 6$ Combine similar terms.

$\frac{-12n}{-12} = \frac{6}{-12}$ Divide both sides by –12.

$n = \frac{1}{-2} = -\frac{1}{2}$ The negative in the denominator is usually shown in front of the fraction.

Check: Try the check on your own. $n = -\frac{1}{2}$ is the solution to the original equation.

e) $4(p - 2) + 3p = -6 + 3p - 2$

$4p - 8 + 3p = -6 + 3p - 2$ Remove the parentheses.

$7p - 8 = -8 + 3p$	Simplify the left side; simplify the right side.
$7p - 8 - \mathbf{3p} = -8 + 3p - \mathbf{3p}$	Subtract $3p$ from both sides.
$4p - 8 = -8$	Combine similar terms.
$4p - 8 + \mathbf{8} = -8 + \mathbf{8}$	Add 8 to both sides.
$4p = 0$	
$\frac{4p}{4} = \frac{0}{4}$	Divide both sides by 4.
$p = 0$	$\frac{0}{4} = 0$. 0 divided by any nonzero number is 0.
Check: $4(p - 2) + 3p = -6 + 3p - 2$	This is the original equation.
$4((0) - 2) + 3(0) \stackrel{?}{=} -6 + 3(0) - 2$	Replace the variable with the proposed solution.
$4(-2) + 0 \stackrel{?}{=} -6 + 0 - 2$	Simplify inside the parentheses.
$-8 + 0 \stackrel{?}{=} -6 + 0 - 2$	Perform multiplication before addition and subtraction.
$-8 = -8$	Both sides are equal, so the solution checks.
$p = 0$	This is the solution to the original equation.

Exercise 2.11

Solve each equation:

a) $\frac{1}{2}a + \frac{1}{4}a - \frac{3}{8}a = -6$

b) $\frac{1}{2}x - \frac{2}{3} = \frac{3}{5}$

c) $0.08y + 2 = 0.2y - 4$

d) $\frac{1}{2}(x + 4) = \frac{1}{6}(x - 3)$

Solution 2.11

a) $\frac{1}{2}a + \frac{1}{4}a - \frac{3}{8}a = -6$	This is the original equation.

$8\left(\frac{1}{2}a+\frac{1}{4}a-\frac{3}{8}a\right)=8(-6)$	The LCD of 2, 4, and 8 is 8.
$8\cdot\frac{1}{2}a+8\cdot\frac{1}{4}a-8\cdot\frac{3}{8}a=-48$	Use the distributive property.
$4a+2a-3a=-48$	Multiply.
$3a=-48$	Combine similar terms.
$\frac{1}{3}\cdot 3a=\frac{1}{3}(-48)$	Multiply by the reciprocal of 3.
$a=-16$	This is the solution.

Compare this to the technique used in Exercise 2.9c. Which was easier for you?

b) $\frac{1}{2}x-\frac{2}{3}=\frac{3}{5}$	This is the original equation.
$30\left(\frac{1}{2}x-\frac{2}{3}\right)=30\left(\frac{3}{5}\right)$	The LCD of 2, 3, and 5 is 30.
$30\cdot\frac{1}{2}x-30\cdot\frac{2}{3}=18$	Use the distributive property.
$15x-20=18$	Multiply.
$15x-20+\mathbf{20}=18+\mathbf{20}$	Add 20 to both sides.
$15x=38$	Simplify each side.
$\frac{1}{15}(15x)=\frac{1}{15}(38)$	Multiply by the reciprocal of 15.
$x=\frac{38}{15}$	This is the solution.
c) $0.08y+2=0.2y-4$	This is the original equation.

$100(0.08y + 2) = 100(0.2y - 4)$	To move the decimal point two places, multiply by 100.
$8y + 200 = 20y - 400$	Use the distributive property.
$8y + 200 - \mathbf{200} = 20y - 400 - \mathbf{200}$	Subtract 200 from both sides.
$8y = 20y - 600$	Combine similar terms.
$8y - \mathbf{20y} = 20y - 600 - \mathbf{20y}$	Subtract $20y$ from both sides.
$-12y = -600$	Combine similar terms.
$\frac{-12y}{-12} = \frac{-600}{-12}$	Divide both sides by -12.
$y = 50$	This is the solution.

d) $\frac{1}{2}(x+4) = \frac{1}{6}(x-3)$ This is the original equation.

It will be faster to clear fractions before multiplying to remove the parentheses:

$6\left[\frac{1}{2}(x+4)\right] = 6\left[\frac{1}{6}(x-3)\right]$	Multiply both sides by 6.
$3(x + 4) = 1(x - 3)$	$6 \cdot \frac{1}{2} = 3, 6 \cdot \frac{1}{6} = 1$
$3x + 12 = x - 3$	Use the distributive property.
$3x + 12 - \mathbf{12} = x - 3 - \mathbf{12}$	Subtract 12 from both sides.
$3x = x - 15$	Combine similar terms.
$3x - \boldsymbol{x} = x - 15 - \boldsymbol{x}$	Subtract x from both sides.
$2x = -15$	Combine similar terms.
$\frac{2x}{2} = -\frac{15}{2}$	Divide both sides by 2.
$x = -\frac{15}{2}$	This is the solution.

2.5 SOLVING AND CHECKING WORD PROBLEMS

Setting up and solving word problems takes practice. This section presents several types of word problems. You should work through these problems more than once to help you perfect your word problem skills.

TRANSLATING WORDS TO SYMBOLS

We have already presented some words dealing with basic symbols and operations. Here are a few other translations:

Addition	**Subtraction**	**Multiplication**	**Division**
Plus	Difference	Product	Quotient
Increased by	Decreased by	Times	Divided by
Sum	Less than	Double (2 times)	Reciprocal
Added to	Minus	Twice (2 times)	Per
Total		Triple (3 times)	
More than		Factor	

Remember: Memorizing these lists is *not* a substitute for reading and understanding the problem. Also, "and" can mean "add," but "Tom and Jerry" does not mean "Tom + Jerry"!

Exercise 2.12

Translate each phrase into a mathematical expression using x as the variable. You may also want to review Section 2.1:

a) 4 more than a number

b) A number increased by 6

c) 5 decreased by a number

d) 2 less than a number

e) Double a number

f) The reciprocal of a number

Solution 2.12

a) $4 + x$

b) $x + 6$

c) $5 - x$

d) $x - 2$ Be careful. "2 less than" means 2 is subtracted from some number, x.

e) $2x$

f) $\frac{1}{x}$

Many word problems can be translated into equations, and then solved. When translating, the following words usually translate to =:

* Is
* Are
* Equals
* Was
* Other forms of the verb "to be"

Exercise 2.13

The sum of twice a number and 5 is 11. Find the number.

Solution 2.13

Begin by translating the English phrases into symbols. If you let x represent the number, "twice a number and 5 is 11" translates to:

$$2x + 5 = 11$$

Now, solve the equation:

$$2x + 5 - \mathbf{5} = 11 - \mathbf{5}$$ Subtract 5 from both sides.

$$2x = 6$$

$$\frac{2x}{2} = \frac{6}{2}$$ Divide both sides by 2.

$$x = 3$$ This is the proposed solution.

Check: With word problems, you must not only check the proposed solution in the original equation but also check to be sure the answer makes sense.

$2x + 5 = 11$	This is the original equation.
$2(3) + 5 \stackrel{?}{=} 11$	Replace the variable with the proposed solution.
$6 + 5 \stackrel{?}{=} 11$	Multiply before adding.
$11 = 11$	Both sides are equal, so the solution checks.

The sum of twice 3 and 5 does equal 11, so the solution makes sense.

Exercise 2.14

If 4 is subtracted from the product of 5 and a number, the result is 26. Find the number.

Solution 2.14

Product of 5 and a number is $5x$. 4 is subtracted from the product is $5x - 4$. The result is 26 provides the = 26 to give:

$$5x - 4 = 26$$

$$5x - 4 + \mathbf{4} = 26 + \mathbf{4}$$ Add 4 to both sides.

$$5x = 30$$

$$\frac{5x}{5} = \frac{30}{5}$$ Divide both sides by 5.

$$x = 6$$

Check: Try the check on your own.

CONSECUTIVE INTEGERS

Number problems may also deal with types of integers, such as consecutive integers, consecutive even integers, and consecutive odd integers. Study these examples:

Term	Using Numbers	Using Variables
Consecutive integers	4, 5, 6, 7	$x, x + 1, x + 2, x + 3$
Consecutive even integers	2, 4, 6, 8	$x, x + 2, x + 4, x + 6$
Consecutive odd integers	5, 7, 9, 11	$x, x + 2, x + 4, x + 6$

Exercise 2.15

Three times the larger of two consecutive odd integers is 17 more than the sum of the two odd integers. Find the integers.

Solution 2.15

Let x = the smaller odd integer.

Then $x + 2$ = the larger odd integer.

Three times the larger is 17 more than the sum:

$3(x + 2) = 17 + (x) + (x + 2)$	Translate.
$3x + 6 = 17 + x + x + 2$	Remove the parentheses, using the distributive property.
$3x + 6 = 19 + 2x$	Simplify each side of the equation.
$3x + 6 - \mathbf{2x} = 19 + 2x - \mathbf{2x}$	Subtract $2x$ from both sides.
$x + 6 = 19$	
$x + 6 - \mathbf{6} = 19 - \mathbf{6}$	Subtract 6 from both sides.
$x = 13$	This is the proposed solution.
And, $x + 2 = 15$	

Check:

$3(x + 2) = 17 + (x) + (x + 2)$	This is the original equation.
$3[(13) + 2] \stackrel{?}{=} 17 + (13) + [(13) + 2]$	Replace the variable with the proposed solution.
$3(15) \stackrel{?}{=} 17 + 13 + (15)$	Simplify inside the parentheses first.
$45 = 45$	Both sides are equal, so the solution checks.

If the integers are 13 and 15, three times 15 is 17 more than their sum (28), so the solution makes sense.

AGE PROBLEMS

Consider this table for several different ages:

Age Now	Age 3 Years Ago	Age 6 Years from Now
38	35	44
21	18	27
24	21	30
x	$x - 3$	$x + 6$

Now, use the idea of this table to help you solve the following exercise:

Exercise 2.16

Joan is 16 years older than Lisa. In 5 years, the sum of their ages will be 70. Find the age of each now.

Solution 2.16

Let x = Lisa's age now.

Then $x + 16$ = Joan's age now.

A table may help you organize this data:

	Age Now	Age 5 Years from Now
Lisa	x	$x + 5$
Joan	$x + 16$	$x + 16 + 5$

The sum of their ages will be 70.

$(x + 5) + (x + 16 + 5) = 70$ Translate.

$2x + 26 = 70$ Simplify the left side.

$2x + 26 - \mathbf{26} = 70 - \mathbf{26}$ Subtract 26 from both sides.

$2x = 44$

$\frac{2x}{2} = \frac{44}{2}$ Divide both sides by 2.

$x = 22$ This is the proposed solution.

And, $x + 16 = 38$

The proposed solution checks in the original equation. In 5 years, Lisa will be 27 and Joan will be 43. Because $27 + 43 = 70$, the solution also makes sense.

Exercise 2.17

Bill is twice as old as Emily. Five years ago, the sum of their ages was 32. How old are they now?

Solution 2.17

Let x = Emily's age now.

Then $2x$ = Bill's age now.

	Age Now	Age 5 Years Ago
Emily	x	$x - 5$
Bill	$2x$	$2x - 5$

The sum of their ages was 32.

$(x - 5) + (2x - 5) = 32$	Translate.
$3x - 10 = 32$	Simplify left side of equation.
$3x - 10 + \mathbf{10} = 32 + \mathbf{10}$	Add 10 to both sides.
$3x = 42$	
$\frac{3x}{3} = \frac{42}{3}$	Divide both sides by 3.
$x = 14$	Proposed solution.
And, $2x = 28$	

Check:

$(x - 5) + (2x - 5) = 32$	Original equation.
$((14) - 5) + (2(14) - 5) \stackrel{?}{=} 32$	Replace the variable with its proposed solution.
$(9) + (28 - 5) \stackrel{?}{=} 32$	Simplify inside parentheses first.
$9 + 23 \stackrel{?}{=} 32$	
$32 = 32$	Both sides are equal, so the solution checks.

If Emily is 14 and Bill is 28, five years ago, Emily was 9 and Bill was 23. Because the sum of 9 and 23 is 32, the solution makes sense.

GEOMETRY PROBLEMS

There are several geometry formulas you should recall (read that as "memorize" if you don't already know them).

Shape	Perimeter
Square	$4s$
Rectangle	$2L + 2W$
Triangle	$a + b + c$

There are several types of problems that involve angles. A few things to know:

* The sum of the measures of the interior angles of a triangle is 180°.
* The angle measure of a straight line is 180°.
* The measure of a right angle is 90°.

* Complementary angles are two angles whose sum measures 90°.
* Supplementary angles are two angles whose sum measures 180°.

Exercise 2.18

The length of a rectangle is 4 feet more than twice its width. The perimeter is 44 feet. Find the length and width.

Solution 2.18

We know the least about the width, so let W = width. Then,

$L = 4 + 2W$	The length is 4 more than twice the width.
$2W + 2(4 + 2W) = 44$	Set up the formula for perimeter.
$2W + 8 + 4W = 44$	Use the distributive property.
$6W + 8 = 44$	Combine similar terms.
$6W + 8 - \mathbf{8} = 44 - \mathbf{8}$	Subtract 8 from both sides.
$6W = 36$	Combine similar terms.
$\frac{6W}{6} = \frac{36}{6}$	Divide both sides by 6.
$W = 6$	

Now, find the length, using $L = 4 + 2W$:

$L = 4 + 2(6)$	Substitute $W = 6$.
$L = 4 + 12 = 16$	Multiply, and then add.

Check:

$P = 2L + 2W$	This is the formula for the perimeter of a rectangle.
$44 = 2(16) + 2(6)$	Substitute for L and W.
$44 = 44$	The solution checks.

Exercise 2.19

A triangle's second side is one-half the length of the first side, and the third side is four times as long as the first. If the perimeter is 110 meters, what is the length of the third side?

Solution 2.19

You know the least about the first side, so use the following:

x = length of first side $\frac{1}{2}x$ = length of second side $4x$ = length of third side	Translate.
$P = a + b + c$	Set up the formula for the perimeter.
$110 = x + \frac{1}{2}x + 4x$	Use this, given that the perimeter is 110 meters.
$2(110) = 2\left(x + \frac{1}{2}x + 4x\right)$	Clear the fractions by multiplying both sides by 2.
$220 = 2x + x + 8x$	Use the distributive property.
$220 = 11x$	Combine similar terms.
$\frac{220}{11} = \frac{11x}{11}$	Divide both sides by 11.
$20 = x$	

Recall that x is the length of the first side, but you are asked for the length of the third side:

$4x$ = length of third side

$4(20)$ = 80 meters = length of third side.

Check:

$x + \frac{1}{2}x + 4x = 110$	This is the original equation.
$20 + \frac{1}{2}(20) + 4(20) = 110$	Substitute $x = 20$.
$20 + 10 + 80 = 110$	The solution checks.

COIN PROBLEMS

When you solve coin problems, keep in mind the relationship between the number of coins and the value of the coins. For example:

Number of Coins	Value
5 dimes	5(0.10) = \$0.50
6 quarters	6(0.25) = \$1.50
3 nickels	3(0.05) = \$0.15

Exercise 2.20

A child's bank contains twice as many nickels as dimes. The total value of the coins is \$1.40. How many dimes are in the bank?

Solution 2.20

You know the least about the number of dimes, so let x = the number of dimes.

$2x$ = the number of nickels — There are twice as many nickels as dimes.

$2x(0.05) + x(0.10) = 1.40$ — This is the value of the coins.

$100[2x(0.05) + x(0.10)] = 100[1.40]$ — Multiply by 100 to clear decimals.

$2x(5) + x(10) = 140$ — Use the distributive property.

$10x + 10x = 140$ — Multiply.

$20x = 140$ — Combine similar terms.

$\frac{20x}{20} = \frac{140}{20}$ — Divide both sides by 20.

$x = 7$

This means there are 7 dimes and 2(7) = 14 nickels.

Check:

7 dimes are worth 7(\$0.10) = \$0.70

14 nickels are worth 14(\$0.05) = \$0.70

Total value = \$0.70 + \$0.70 = \$1.40

2.6 FORMULAS

Some formulas you may be familiar with are presented in the following table:

Formula	Formula Is Used For	Description of the Formula
$A = LW$	Area of a rectangle	The **Area** of a rectangle equals the **Length** times the **Width**.
$A = \frac{1}{2}bh$	Area of a triangle	The **Area** of a triangle equals $\frac{1}{2}$ times the length of the **base** times the **height**.
$A = \pi r^2$	Area of a circle	The **Area** of a circle equals π times the **radius** squared.
$C = 2\pi r$	Circumference of a circle	The **Circumference** of a circle equals 2 times π times the **radius**.

Notice that we often use the first letter of an important word as the variable in the formula (A for area, L for length, and so on). In the circle formulas above, π stands for the irrational number pi. You may be given a fraction or decimal approximation to replace π in some cases.

USING FORMULAS TO SOLVE WORD PROBLEMS

You can find the value of one of the variables if you know the values of the other variables in a formula.

Exercise 2.21

Find the length of a rectangle that has perimeter 20 feet and width 4 feet.

Solution 2.21

Let L = the length of the rectangle and W = the width of the rectangle. Then:

$P = 2L + 2W$ Write the formula.

$(20) = 2L + 2(4)$ Substitute the given values.

$20 = 2L + 8$ Simplify right side of an equation.

$20 - \mathbf{8} = 2L + 8 - \mathbf{8}$ Subtract 8 from both sides.

$12 = 2L$

$\frac{12}{2} = \frac{2L}{2}$ Divide both sides by 2.

$6 = L$ This is the proposed solution.

Check:

$P = 2L + 2W$	This is the original equation.
$20 \stackrel{?}{=} 2(6) + 2(4)$	Replace L with the proposed solution and W with its given measurement.
$20 \stackrel{?}{=} 12 + 8$	Multiply before adding.
$20 = 20$	Both sides are equal, so the solution checks.

Exercise 2.22

The volume of a circular cone is $V = \frac{1}{3}\pi r^2 h,$ where r is the radius of the base and h is the height of the cone. If the volume is 352 cubic inches and the radius is 2 inches, find the height of the cone. Use $\frac{22}{7}$ for π.

Solution 2.22

$V = \frac{1}{3}\pi r^2 h$	Write the formula.
$(352) = \frac{1}{3}\left(\frac{22}{7}\right)(2)^2 h$	Substitute the given values.
$352 = \frac{1}{3}\left(\frac{22}{7}\right)4h$	Simplify exponents.
$352 = \left(\frac{88}{21}\right)h$	Simplify the right side.
$\frac{21}{88}\cdot 352 = \frac{21}{88}\cdot\left(\frac{88}{21}\right)h$	Multiply both sides by the reciprocal of the coefficient of h.
$84 = h$	

Thus, the height of the cone is 84 inches.

2.7 SOLVING A FORMULA FOR ONE OF ITS VARIABLES

You may also need to solve a formula for one of the variables. Follow the rules for solving a linear equation to get the desired variable alone on one side of the equation.

Exercise 2.23

Solve $A = LW$ for L.

Solution 2.23

$A = LW$	Write the formula.
$\frac{A}{W} = \frac{LW}{W}$	The coefficient of L is W. Divide both sides by W.
$\frac{A}{W} = L$	
$L = \frac{A}{W}$	The formula is now solved for L.

Exercise 2.24

Solve $y = mx + b$ for x.

Solution 2.24

$y = mx + b$	Write the formula.
$y - b = mx + b - b$	Subtract b from both sides of the equation.
$y - b = mx$	
$\frac{y-b}{m} = \frac{mx}{m}$	The coefficient of x is m. Divide both sides by m.
$\frac{y-b}{m} = x$	
$x = \frac{y-b}{m}$	The formula is now solved for x.

Exercise 2.25

Solve $P = 2L + 2W$ for W.

Solution 2.25

$P = 2L + 2W$	Write the formula.
$P - 2L = 2L + 2W - 2L$	Subtract $2L$ from both sides.
$P - 2L = 2W$	

$$\frac{P-2L}{2}=\frac{2W}{2}$$ The coefficient of W is 2. Divide both sides by 2.

$$\frac{P-2L}{2}=W$$

$$W=\frac{P-2L}{2}$$ The formula is now solved for W.

2.8 SOLVING LINEAR INEQUALITIES

A linear inequality is similar to a linear equation, except that the equality symbol (=) is replaced by one of the four inequality symbols ($<$, $>$, $\leq$, $\geq$). The solution of a linear inequality is often graphed on a number line. We begin by graphing inequalities like $x < 4$ and $x > 2$.

GRAPHING INEQUALITIES

To graph an inequality in the form $x < a$, do the following:

1. Put an open circle on a.
2. Shade to the left of a, using an arrow to indicate that the shading continues.

To graph an inequality in the form $x > a$, do the following:

1. Put an open circle on a.
2. Shade to the right of a, using an arrow to indicate that the shading continues.

If the inequality symbol is $\leq$ or $\geq$, use a closed (darkened) circle on a.

Exercise 2.26

Graph:

a) $x < 4$

b) $x \geq -1$

Solution 2.26

a) To graph $x < 4$, begin by putting an open circle on 4. Next, shade to the *left* of 4:

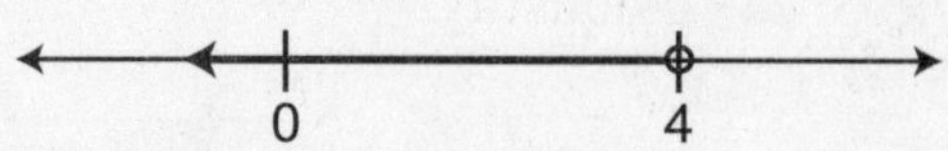

b) To graph $x \geq -1$, begin by putting a closed circle on -1. Next, shade to the *right* of -1:

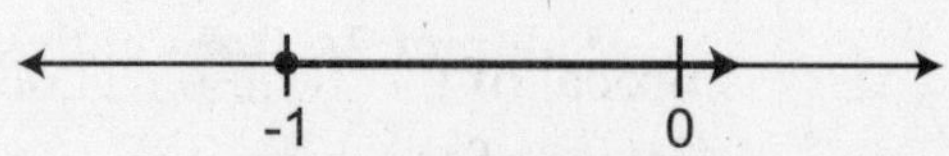

Exercise 2.27

Graph:

a) $2 < x$

b) $-3 \geq x$

Solution 2.27

a) First rewrite the inequality to get the variable on the left. $2 < x$ is equivalent to $x > 2$. (Remember to keep the inequality symbol "pointing" at the 2.) To graph $x > 2$, put an open circle on 2. Next, shade to the *right* of 2:

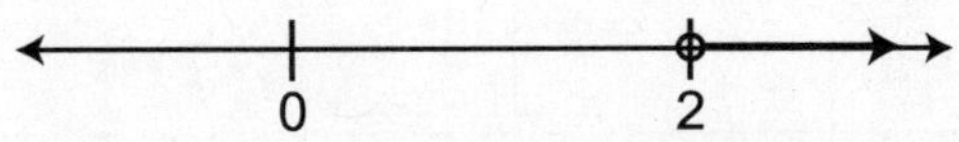

b) First rewrite the inequality to get the variable on the left. $-3 \geq x$ is equivalent to $x \leq -3$. (Remember to keep the inequality symbol "pointing" at the x.) To graph $x \leq -3$, put a closed circle on -3. Next, shade to the *left* of -3:

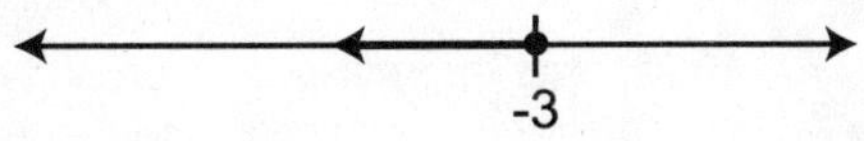

SOLVING LINEAR INEQUALITIES

You solve linear inequalities almost the same way you solve linear equations. To solve a linear inequality, do the following:

1. Simplify the left side of the inequality by removing parentheses, and adding and subtracting similar terms.

2. Simplify the right side of the inequality by removing parentheses, and adding and subtracting similar terms.

3. Clear fractions by multiplying both sides by the LCD. Clear decimals by multiplying both sides by an appropriate power of 10. (Keep in mind Step 6 below.)

4. Use addition to get the variable terms together on one side of the inequality and the numbers on the other side of the inequality.

5. Multiply both sides by the reciprocal of the numerical coefficient of the variable (or divide both sides by the numerical coefficient of the variable).

6. If you multiply (or divide) by a negative number, you must reverse the inequality symbol.

The exception to the rules for linear equations occurs in the last step. If you multiply or divide by a negative number, make one of the following changes:

Multiplying or Dividing by a Negative	
Change	**To**
$>$	$<$
$<$	$>$
$\geq$	$\leq$
$\leq$	$\geq$

Exercise 2.28

Solve and graph:

a) $3x < 12$

b) $-2x \leq 2$

Solution 2.28

a) $3x < 12$	Original inequality.
$\frac{3x}{3} < \frac{12}{3}$	Divide both sides by 3.
$x < 4$	This is the solution.

Put an open circle on 4.

Next, shade to the *left* of 4:

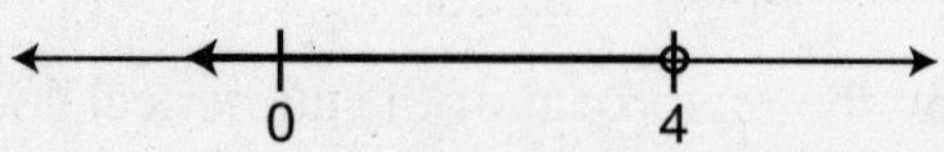

b) $-2x \le 2$

$\frac{-2x}{-2} \ge \frac{2}{-2}$ — Divide both sides by -2; reverse the inequality symbol.

$x \ge -1$ — This is the solution.

Put a closed circle on -1.

Next, shade to the *right* of -1:

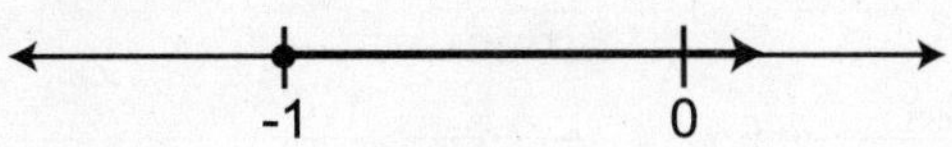

Exercise 2.29

Solve and graph:

a) $4(x - 6) \le -8$

b) $5 + 2x < 4(x + 1) - 7 + 2x$

c) $-\frac{2}{5}a + 9 \ge \frac{1}{2}a$

d) $-4 - \frac{2}{3}y < 2$

e) $-6.2x - 4.4 \le -2.5x + 3$

Solution 2.29

a) $4(x - 6) \le -8$ — Original inequality.

$4x - 24 \le -8$ — Remove the parentheses, using the distributive property.

$4x - 24 + \mathbf{24} \le -8 + \mathbf{24}$ — Add 24 to both sides.

$4x \le 16$ — Combine similar terms.

$\frac{4x}{4} \le \frac{16}{4}$ — Divide both sides by 4.

$x \le 4$ — This is the solution.

Put a closed circle on 4.

Next, shade to the *left* of 4:

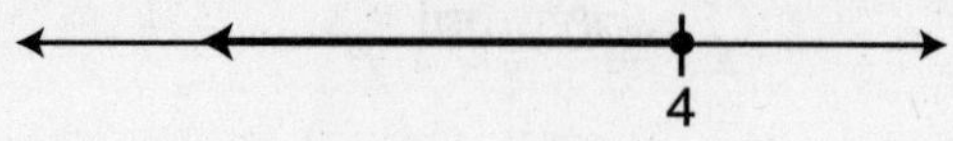

b) $5 + 2x < 4(x + 1) - 7 + 2x$

$5 + 2x < 4x + 4 - 7 + 2x$	Remove the parentheses by using the distributive property.
$5 + 2x < 6x - 3$	Combine similar terms.
$5 + 2x - \mathbf{6x} < 6x - 3 - \mathbf{6x}$	Subtract $6x$ from both sides.
$5 - 4x < -3$	Combine similar terms.
$5 - 4x - \mathbf{5} < -3 - \mathbf{5}$	Subtract 5 from both sides.
$-4x < -8$	Simplify.
$\frac{-4x}{-4} > \frac{-8}{-4}$	Divide both sides by -4.
$x > 2$	This is the solution.

Put an open circle on 2.

Next, shade to the *right* of 2.

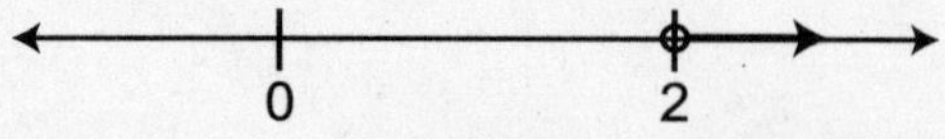

c) $-\frac{2}{5}a + 9 \geq \frac{1}{2}a$	This is the original inequality.
$10\left(-\frac{2}{5}a + 9\right) \geq 10\left(\frac{1}{2}a\right)$	Clear the fractions by multiplying both sides by 10.
$10\left(-\frac{2}{5}a\right) + 10(9) \geq 10\left(\frac{1}{2}a\right)$	Use the distributive property.

$-4a + 90 \geq 5a$

$-4a + 90 - \mathbf{90} \geq 5a - \mathbf{90}$ Subtract 90 from both sides.

$-4a - \mathbf{5a} \geq 5a - 90 - \mathbf{5a}$ Subtract $5a$ from both sides.

$-9a \geq -90$ Combine similar terms.

$\frac{-9a}{-9} \leq \frac{-90}{-9}$ Divide both sides by –9. Reverse the inequality symbol.

$a \leq 10$

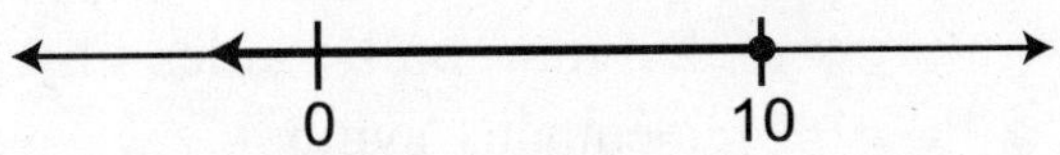

d) $-4 - \frac{2}{3}y < 2$ This is the original inequality.

$3\left(-4 - \frac{2}{3}y\right) < 3(2)$ Clear fractions by multiplying both sides by 3.

$3(-4) - 3\left(\frac{2}{3}y\right) < 6$ Use the distributive property.

$-12 - 2y < 6$ Multiply.

$-12 - 2y + \mathbf{12} < 6 + \mathbf{12}$ Add 12 to both sides.

$-2y < 18$ Combine similar terms.

$\frac{-2y}{-2} > \frac{18}{-2}$ Divide both sides by –2. Reverse the inequality symbol.

$y > -9$

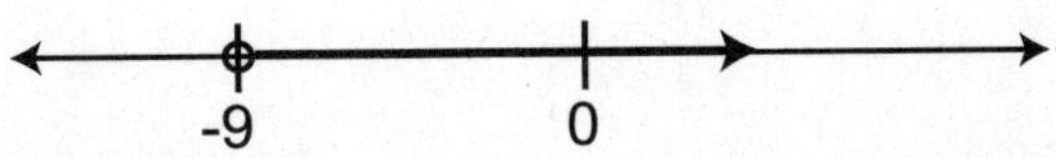

e) $-6.2x - 4.4 \le -2.5x + 3$ — This is the original inequality.

$10(-6.2x - 4.4) \le 10(-2.5x + 3)$ — Clear decimals by multiplying both sides by 10.

$-62x - 44 \le -25x + 30$ — Use the distributive property.

$-62x - 44 + \mathbf{44} \le -25x + 30 + \mathbf{44}$ — Add 44 to both sides.

$-62x \le -25x + 74$ — Combine similar terms.

$-62x + \mathbf{25x} \le -25x + 74 + \mathbf{25x}$ — Add 25x to both sides.

$-37x \le 74$ — Combine similar terms.

$\dfrac{-37x}{-37} \ge \dfrac{74}{-37}$ — Divide both sides by -37. Reverse the inequality symbol.

$x \ge -2$

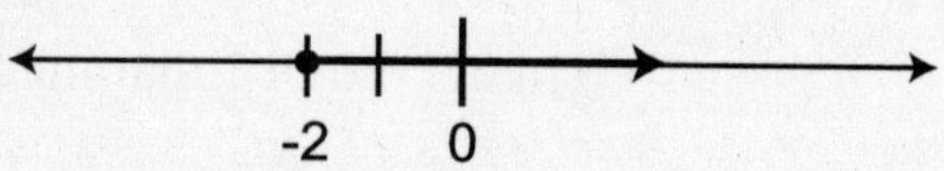

SOLVING WORD PROBLEMS WITH LINEAR INEQUALITIES

You may need to solve word problems that involve inequalities. Some language translations are as follows; x represents "the number."

* The number is at most: $x \le$
* The number is at least: $x \ge$
* The number is more than: $x >$
* The number is less than: $x <$

Exercise 2.30

The sum of two consecutive odd integers is at most 32. What are the possibilities for the first of the two integers?

Solution 2.30

Let x = first odd integer.

Then let $x + 2$ = next consecutive odd integer.

The sum is at most 32, so

$(x) + (x + 2) \le 32$	Translate.
$x + x + 2 \le 32$	
$2x + 2 \le 32$	Combine similar terms.
$2x + 2 - \mathbf{2} \le 32 - \mathbf{2}$	Subtract 2 from both sides
$2x \le 30$	
$\frac{2x}{2} \le \frac{30}{2}$	Divide both sides by 2.
$x \le 15$	This is the solution.

The first integer must be an odd integer less than or equal to 15, and the next consecutive odd integer must be less than or equal to 17. Then their sum must be less than or equal to $15 + 17 = 32$. Thus, $x \le 15$ checks.

TEST YOURSELF

1) Simplify by combining similar or like terms:

(a) $2uv - 3uv$

(b) $-7y^3 + y^3$

(c) $11x + 5x - 6 + 4$

(d) $6x^2 - 1 + 3x - 7x^2 - 9x + 8$

(e) $p + 9p^2$

2) Simplify:

(a) $7(3p + 2)$

(b) $-4(6x - 8)$

(c) $2(9t + 4s) - 3(2t - 8s)$

(d) $6 + 2(4a - 3)$

(e) $5(1 - 6a) + 7(2 - 3b)$

3) Evaluate each expression using the given value of the variable(s):

(a) $7y - 5$; $y = 2$

(b) $4x + 6$; $x = -4$

(c) $(6p - 3q)(4)$; $p = 1$, $q = -3$

(d) $t^2 - 3t + 9$; $t = 8$

(e) $(4p - 3q)(p^2 - 4q + p)$; $p = 2$, $q = -1$

4) Translate each phrase into an algebraic expression, using x as the unknown number. Then find the value of the expression when $x = -4$:

(a) The sum of a number and 2, increased by 8.

(b) The difference of a number and 3, decreased by –6.

(c) Twice a number, decreased by 5.

(d) Six, subtracted from the sum of a number and 1.

(e) The product of a number and –2, minus the sum of a number and 4.

(f) The sum of a number and 2, subtracted from –8.

5) Solve and check each equation:

(a) $x - 4 = 10$

(b) $y - 3 = -7$

(c) $t - \frac{1}{4} = \frac{2}{5}$

(d) $t - 2.3 = -4.7$

6) Solve and check each equation:

(a) $2x + 6 = 4$

(b) $x - 3 = 7$

(c) $y + 9 = 12$

(d) $8(y - 2) = 16$

(e) $6m + 5 = 1$

7) Solve and check each equation:

(a) $p + 4 = 7$

(b) $6t = 5t - 2$

(c) $a - 3 = -7 + 2a$

(d) $8s + 3s - 9 = -16s + 3 + 3s$

(e) $5m + 2(3m - 1) = 9$

(f) $3y + \frac{1}{4} = 2y + \frac{2}{3}$

(g) $5x - 6.3 = 4x - 8.2$

8) Solve:

(a) $\frac{1}{4}y = 8$

(b) $-2x = 14$

(c) $\frac{a}{7} = 21$

(d) $-\frac{2}{3}t = 24$

(e) $14b = 84$

9) Solve each equation:

(a) $5x - 7x = -8$

(b) $2 - 7 = -4y - 6y$

(c) $4a - 5a - 6a = 14$

(d) $22 - 14 = \frac{1}{6}x + \frac{2}{3}x - \frac{1}{2}x$

10) Solve and check each equation:

(a) $4(6 - n) - 3(8n + 9) = 12$

(b) $2x - 5 + 4x = 3 - 7x + 9$

(c) $3(p + 9) = 9(p + 3)$

(d) $12y(2 - 8) - 12y = 0$

(e) $11t - 2(t + 3) = 4 + t - 2t$

11) Solve each equation:

(a) $\frac{2}{3}n - \frac{1}{4}n = \frac{1}{3} + \frac{1}{2}$

(b) $\frac{4}{5}x + 2 = \frac{1}{3}$

(c) $\frac{1}{3}(2x - 1) = \frac{2}{5}(x + 4)$

(d) $0.2y + 1.4 = 0.3y - 2$

(e) $0.05a + 0.04(20 - a) = 4.2$

12) Translate each phrase into a mathematical expression using x as the number:

(a) The reciprocal of a number

(b) A number increased by 7

(c) Triple a number

(d) 4 more than twice a number

(e) 3 less than a number decreased by 2

13) Find the numbers:

(a) The difference of a number and 5 is 6.

(b) If 8 is added to the product of 2 and a number, the result is 14.

(c) A number is tripled and added to 9. The result is six times the number itself.

(d) A number divided by 5 and added to 20 is the number itself.

(e) The sum of 12 divided by a number and 7 is 10.

14) Solve:

(a) Two times the smallest of three consecutive odd integers is one less than the largest integer. Find the integers.

(b) 6 less than the middle integer of 3 consecutive even integers is 30 less than the sum of the other two integers. Find the integers.

(c) Lisa is 11 years younger than James. In 12 years, James will be twice as old as Lisa is now. How old is James?

(d) Bill is 4 years younger than Beth. In 10 years, the sum of their ages will be 72. Find their ages.

15) Solve:

(a) A triangle has, for the lengths of its sides, three consecutive integers. If the distance around the triangle is 78 centimeters, what is the length of the shortest side?

(b) A triangle's second side is one-fourth the length of its first side, and the third side is three times as long as the first. If the perimeter of the triangle is 68 inches, what is the length of the third side?

(c) One angle is 4 times its complement. Find the measures of the two angles.

(d) One angle is 60° more than twice its supplement. Find the measures of the two angles.

(e) Two angles of a triangle are equal. The third angle is 30° less than either of the other angles. Find the measure of each angle.

16) Solve:

(a) A child's bank contains six times as many dimes as nickels. If the total value of the coins is $1.30, how many dimes are in the bank?

(b) Bill has 8 more dimes than he has quarters. If the total value of his coins is \$2.55, how many dimes does he have?

(c) Kathy Lynn has a collection of nickels, dimes, and quarters totaling \$2.60. She has two fewer dimes than nickels, and $\frac{1}{3}$ the number of nickels are quarters. How many of each coin does she have?

17) Solve:

(a) Find the length of a rectangle that has perimeter 32 feet and width 6 feet.

(b) The volume of a circular cone is $V = \frac{1}{3}\pi r^2 h$. If the volume is 352 cubic inches and the height is 84 inches, find the radius. Use $\pi = \frac{22}{7}$.

(c) The perimeter of an equilateral triangle is 39 meters. Find the length of each side.

(d) If one side of a square is 15 cm less than the perimeter, find the length of a side of the square.

18) Solve for the indicated variable:

(a) $A = \frac{1}{2}bh$ for b

(b) $A = \pi r^2$ for r

(c) $y = mx + b$ for m

(d) $A = LW$ for W

(e) $F = ma$ for m

19) Graph:

(a) $x > 4$

(b) $x \leq 3$

(c) $x < -2$

(d) $x \geq -9$

(e) $x = 5$

20) Solve and graph:

(a) $3(2x - 4) \leq 6$

(b) $-(x - 9) > 3 + x$

(c) $8(2x + 4) \geq 3(5x - 7)$

(d) $2(3 - x) < 2(2x - 3)$

(e) $7x \geq 2(3x + 9)$

21) Solve the following inequalities. Graph the solution set:

(a) $x - \frac{1}{8} > \frac{1}{2}$

(b) $-\frac{7}{8} > -56t$

(c) $1 - \frac{3x}{2} \geq x - 4$

(d) $3.7x \geq 4.5x - 10 - 5.8x$

(e) $1.2x - 3.6 \leq -0.3x + 2.4$

TEST YOURSELF ANSWERS

1) (a) $-uv$
   (b) $-6y^3$
   (c) $16x - 2$
   (d) $-x^2 - 6x + 7$
   (e) $p + 9p^2$

2) (a) $21p + 14$
   (b) $-24x + 32$
   (c) $12t + 32s$
   (d) $8a$
   (e) $19 - 30a - 21b$

3) (a) 9
   (b) -10
   (c) 60
   (d) 49
   (e) 110

4) (a) $(x + 2) + 8$; 6
   (b) $(x - 3) - (-6)$; -1
   (c) $2x - 5$; -13
   (d) $(x + 1) - 6$; -9
   (e) $-2x - (x + 4)$; 8
   (f) $-8 - (x + 2)$; -6

5) (a) 14
   (b) -4
   (c) $\frac{13}{20}$
   (d) -2.4

6) (a) -1
   (b) 10
   (c) 3
   (d) 4
   (e) $-\frac{2}{3}$

7) (a) 3
   (b) -2
   (c) 4
   (d) $\frac{1}{2}$
   (e) 1
   (f) $\frac{5}{12}$
   (g) -1.9

8) (a) 32
   (b) -7
   (c) 147
   (d) -36
   (e) 6

9) (a) 4
   (b) $\frac{1}{2}$
   (c) -2
   (d) 24

10) (a) $-\frac{15}{28}$

(b) $\frac{17}{13}$

(c) 0

(d) 0

(e) 1

11) (a) 2

(b) $-\frac{25}{12}$

(c) $\frac{29}{4}$

(d) 34

(e) 340

12) (a) $\frac{1}{x}$

(b) $x + 7$

(c) $3x$

(d) $2x + 4$

(e) $(x - 2) - 3$

13) (a) 11

(b) 3

(c) 3

(d) 25

(e) 4

14) (a) 3, 5, 7

(b) 22, 24, 26

(c) 34

(d) Beth is 28, Bill is 24

15) (a) 25 cm

(b) 48 in.

(c) 18°, 72°

(d) 40°, 140°

(e) 40°, 70°, 70°

16) (a) 12

(b) 13

(c) 12 nickels, 10 dimes, 4 quarters

17) (a) 10 feet

(b) 2 inches

(c) 13 meters

(d) 5 cm

18) (a) $\frac{2A}{h}$

(b) $\sqrt{\frac{A}{\pi}}$

(c) $\frac{y-b}{x}$

(d) $\frac{A}{L}$

(e) $\frac{F}{a}$

19) a) 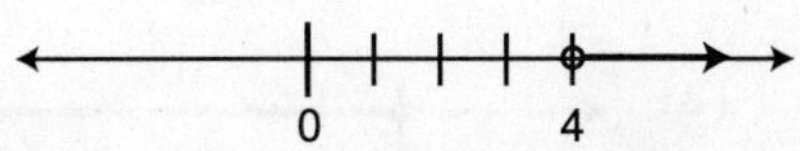

(b) 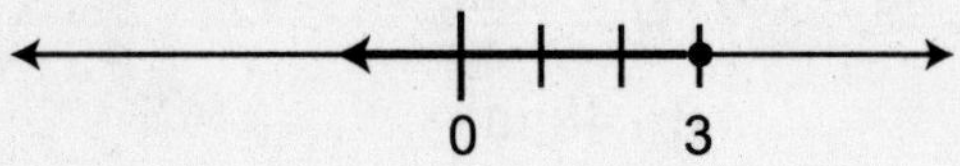

(c) 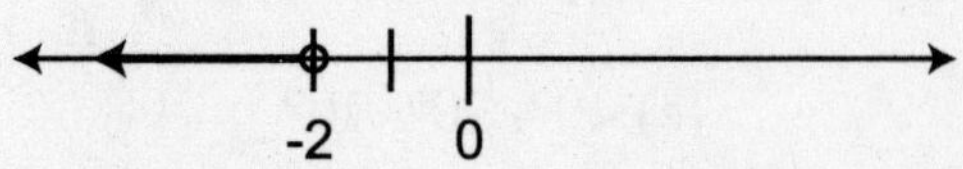

(d) 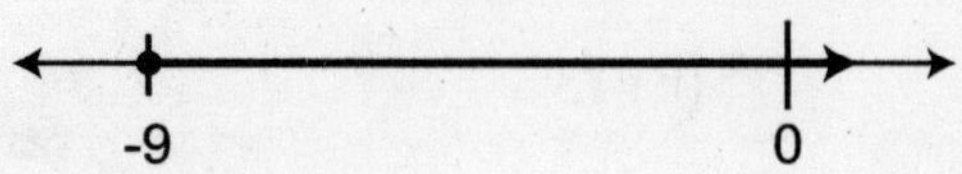

(e) 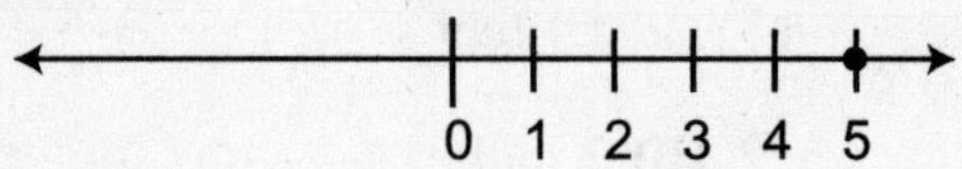

20) (a) 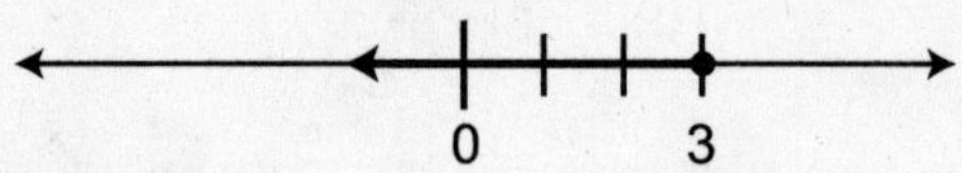

(b) 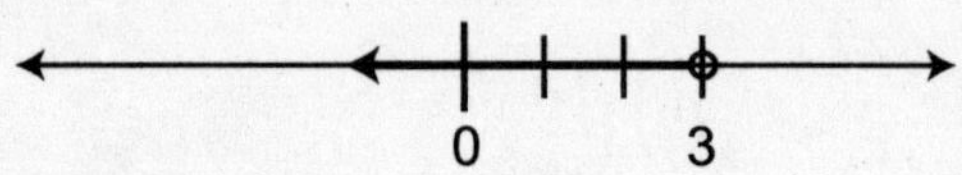

(c) 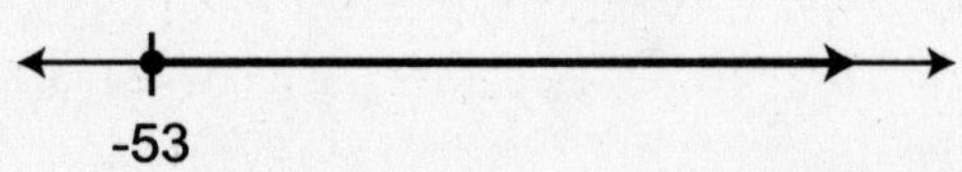

(d) 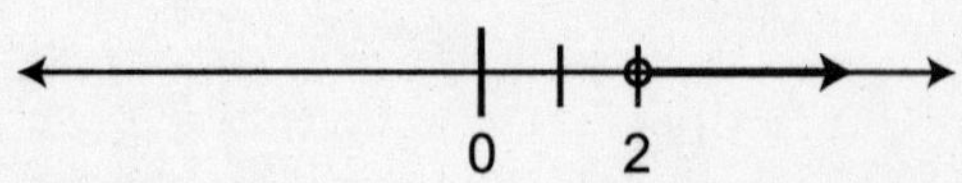

(e) 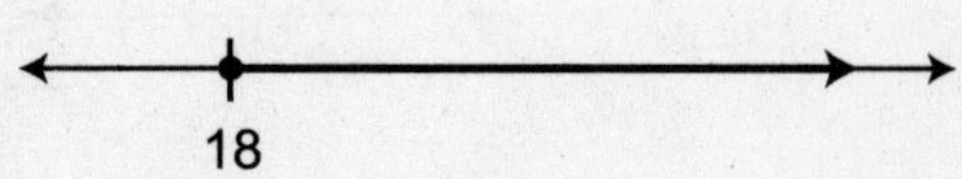

21) (a) $x > \frac{5}{8}$

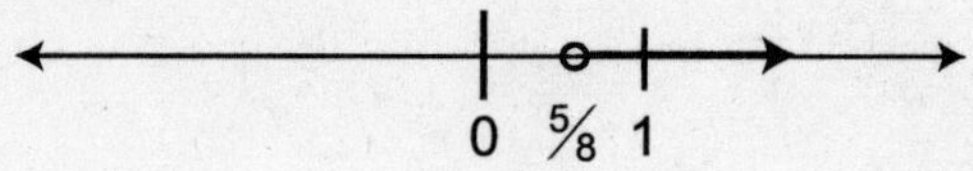

(b) $t > \frac{1}{64}$

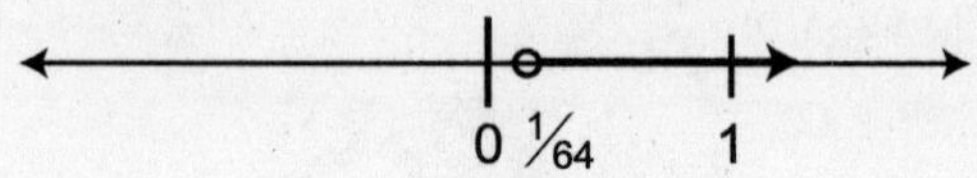

(c) $x \leq 2$

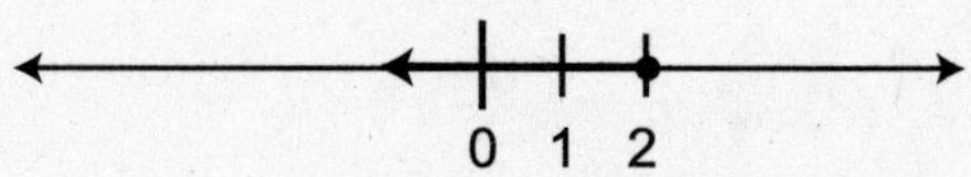

(d) $x \geq -2$

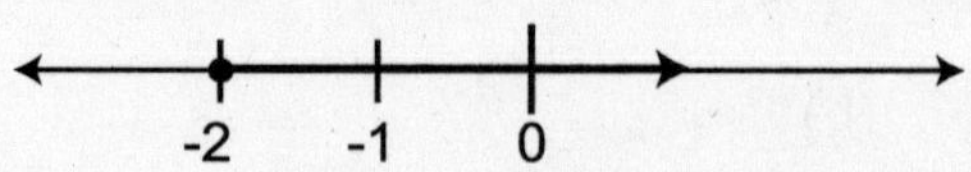

(e) $x \leq 4$

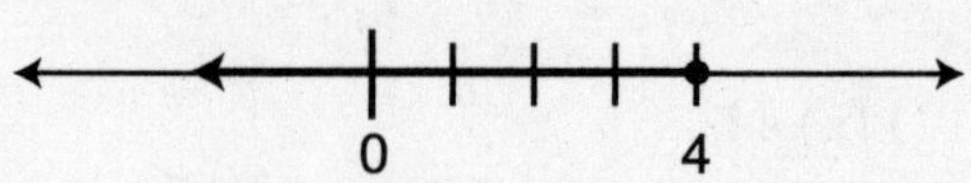

CHAPTER 3

Graphing Linear Equations

The graph of a linear equation provides a visual display of the information contained in the equation. Applications in many fields have data that are represented by linear equations, so you should expect to see the material from this chapter in your future courses.

3.1 SOLUTIONS TO LINEAR EQUATIONS IN TWO VARIABLES

In this chapter, we will solve linear equations in two variables, such as $3x - 2y = 6$ and graph the solutions. A linear equation in two variables is an equation that can be written in the form $ax + by = c$, where a, b, and c are real numbers, and a and b are not both 0. Recall that a solution to an equation like $4x + 6 = 14$ is a number that replaces the variable and makes the equation a true statement. For example, if $x = 2$, $4(2) + 6 = 14$ is a true statement. Solutions to linear equations in two variables will need replacements for both variables. For example, substituting $x = 4$ and $y = 3$ into $3x - 2y = 6$, we have

$3x - 2y = 6$

$3(4) - 2(3) = 6$, which is a true statement.

ORDERED PAIRS

We call $x = 4$ and $y = 3$ a solution to the linear equation in two variables and write it as (4, 3). When a pair of numbers is written in a specific order in parentheses, it is called an **ordered pair.**

COORDINATES

Just as the equations in Chapters 2, 4, and 5 use variables other than x, linear equations in two variables may involve x's and y's, or a's and b's, and so on. The numbers written in an ordered pair are called **coordinates.** Using the example (4, 3), 4 is the x-coordinate and 3 is the y-coordinate, because the equation contained the variables x and y. We generally list the coordinates of an ordered pair in alphabetical order. Consider the following examples:

Linear Equation in Two Variables	Ordered Pair	Example of Solution
$3x + y = 9$	(x, y)	(2, 3) is a solution because $3(2) + (3) = 9$
$x - 5y = 10$	(x, y)	(0, −2) is a solution because $0 - 5(-2) = 10$
$-2x + 5y = 15$	(x, y)	(10, 7) is a solution because $-2(10) + 5(7) = 15$

Exercise 3.1

Determine whether the given ordered pair is a solution of the given equation:

a) $4x + y = 6$; (2, −2)

b) $y = 5x - 1$; (3, 14)

c) $2x - 7y = 6$; (1, 3)

d) $x = 3$; (3, −1)

Solution 3.1

a) $4x + y = 6$; (2, −2) — The ordered pair is in alphabetical order, so $x = 2$ and $y = -2$.

$4(2) + (-2) = 6$ — Substitute $x = 2$, $y = -2$.

$8 + (-2) = 6$ — This is a true statement.

Because this is a true statement, the ordered pair (2, −2) is a solution to $4x + y = 6$.

b) $y = 5x - 1$; (3, 14) — The ordered pair is in alphabetical order, so $x = 3$ and $y = 14$.

$14 = 5(3) - 1$ — Substitute $x = 3$, $y = 14$.

$14 = 15 - 1$ — This is a true statement.

Because this is a true statement, the ordered pair (3, 14) is a solution to $y = 5x - 1$.

c) $2x - 7y = 6$; (1, 3) — The ordered pair is in alphabetical order, so $x = 1$ and $y = 3$.

$2(1) - 7(3) = 6$ — Substitute $x = 1$, $y = 3$.

$2 - 21 = 6$ — Multiply before subtracting.

$-19 = 6$ — This is a false statement.

Because this is a false statement, the ordered pair (1, 3) is *not* a solution to $2x - 7y = 6$.

d) $x = 3$; (3, −1) — The ordered pair is in alphabetical order, so $x = 3$ and $y = -1$.

$x + 0y = 3$ — There is no y written in the given equation, which means the equation can be written $x + 0y = 3$.

$(3) + 0(-1) = 3$ — Substitute $x = 3$ and $y = -1$.

$3 + 0 = 3$ — This is a true statement.

Because this is a true statement, the ordered pair (3, −1) is a solution to $x = 3$.

To solve linear equations with two variables, do the following:
Given one coordinate, find the solution by

1. Substituting the given coordinate.
2. Solving the resulting equation.

Exercise 3.2

Complete the given ordered pairs for the equation $3x - 2y = 6$:

a) (6,)

b) (, 0)

c) (−3,)

Solution 3.2

a) $3x - 2y = 6$

$3(6) - 2y = 6$ — Substitute $x = 6$.

$18 - 2y = 6$ — Solve for y.

$18 - 2y - \mathbf{18} = 6 - \mathbf{18}$ — Subtract 18 from each side of the equation.

$-2y = -12$ — Combine similar terms.

$\frac{-2y}{-2} = \frac{-12}{-2}$	Divide each side by –2.
$y = 6$	Divide.
(6, 6)	This is the solution.
Check:	
$3x - 2y = 6$	This is the original equation.
$3(6) - 2(6) \stackrel{?}{=}$	Substitute $x = 6$ and $y = 6$ into the original equation. The question mark means we are checking.
$18 - 12 \stackrel{?}{=} 6$	Multiply before subtracting.
$6 = 6$	(6, 6) checks.
b) $3x - 2y = 6$	
$3x - 2(0) = 6$	Substitute $y = 0$.
$3x - 0 = 6$	Solve for x.
$3x = 6$	Simplify.
$\frac{3x}{3} = \frac{6}{3}$	Divide each side by 3.
$x = 2$	Divide.
(2, 0)	This is the solution.
Check:	
$3x - 2y = 6$	This is the original equation.
$3(2) - 2(0) \stackrel{?}{=} 6$	Substitute $x = 2$ and $y = 0$ into the original equation.
$6 - 0 \stackrel{?}{=} 6$	Multiply before subtracting.
$6 = 6$	(2, 0) checks.
c) $3x - 2y = 6$	
$3(-3) - 2y = 6$	Substitute $x = -3$.
$-9 - 2y + \mathbf{9} = 6 + \mathbf{9}$	Add 9 to both sides of the equation.
$-2y = 15$	Solve for y.
$\frac{-2y}{-2} = \frac{15}{-2}$	Divide each side by –2.

$y = \frac{-15}{2}$

$\left(-3, -\frac{15}{2}\right)$ This is the solution.

Check:

$3x - 2y = 6$ This is the original equation.

$3(-3) - 2\left(-\frac{15}{2}\right) \stackrel{?}{=} 6$ Substitute $x = -3$ and $y = \frac{-15}{2}$ into the original equation.

$-9 + 15 \stackrel{?}{=} 6$ Multiply before adding.

$6 = 6$ $\left(-3, -\frac{15}{2}\right)$ checks.

Exercise 3.3

Complete the given ordered pairs for the equation $y = 4$:

a) (2,)

b) (−3,)

c) (0,)

Solution 3.3

Because y must equal 4, all the missing y-coordinates are 4. Thus, the required solutions are

a) (2, 4)

b) (−3, 4)

c) (0, 4)

Exercise 3.4

Complete the given ordered pairs for the equation $x = -6$:

a) (, 3)

b) (, −2)

c) (, 0)

Solution 3.4

Because x must equal −6, all the missing x-coordinates are −6. Thus, the required solutions are

a) (−6, 3)
b) (−6, −2)
c) (−6, 0)

The Coordinate System

To graph ordered pairs, we use a rectangular coordinate system made up of two number lines: the horizontal number line is called the ***x*-axis,** and the vertical number line is called the ***y*-axis**.

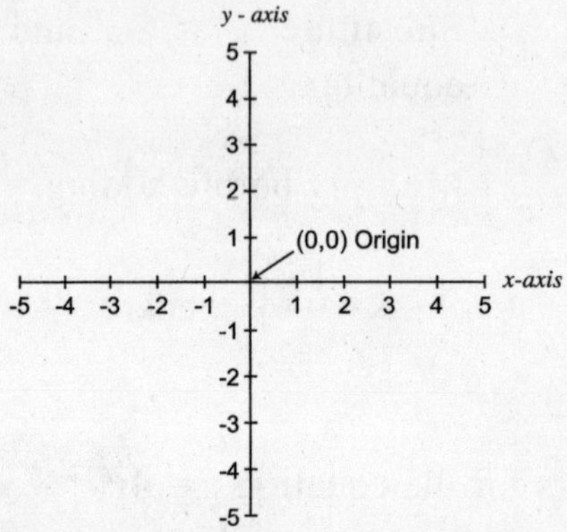

On a number line, positive numbers are to the right of 0, and negative numbers are to the left of 0. Notice that the *x*-axis is arranged in the same way. On the *y*-axis, positive numbers are written up the axis, and negative numbers are written down the axis. Where the two axes (the plural of axis) intersect, the ***x*-coordinate** equals 0 and the ***y*-coordinate** equals 0. The point (0,0) is called the **origin.**

To plot the ordered pair (*x, y*), do the following:

1. Place your pencil point at the origin (0, 0).
2. Go right x units if $x > 0$ or left x units if $x < 0$.
3. Go up y units if $y > 0$ or down y units if $y < 0$.
4. Insert a point where your movement ended, and label it with (*x, y*).

Exercise 3.5

Plot the points (0, 0), (2, 4), (−2, 4), (−2, −4), (2, −4), (0, 4) and (2, 0).

Solution 3.5

(0, 0) is at the intersection of the axes.

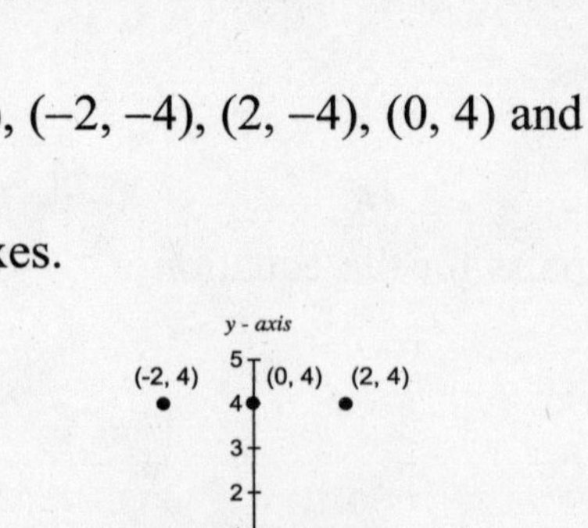

(2, 4) is 2 units to the right and 4 units up from the origin (0, 0).

(–2, 4) is 2 units to the left and 4 units up from the origin (0, 0).

(–2, –4) is 2 units to the left and 4 units down from the origin.

(2, –4) is 2 units to the right and 4 units down from the origin.

(0, 4) is 0 units left or right and 4 units up from the origin.

(2, 0) is 2 units to the right and 0 units up or down from the origin.

3.2 GRAPHING LINEAR EQUATIONS

You can solve $2x - 4 = 10$ as follows:

$2x - 4 + 4 = 10 + 4$ — Add 4 to both sides of the equation.

$2x = 14$ — Combine similar terms.

$\frac{1}{2}(2x) = \frac{1}{2}(14)$ — Multiply each side by $\frac{1}{2}$ (or divide by 2).

$x = 7$

You can graph the solution on a number line:

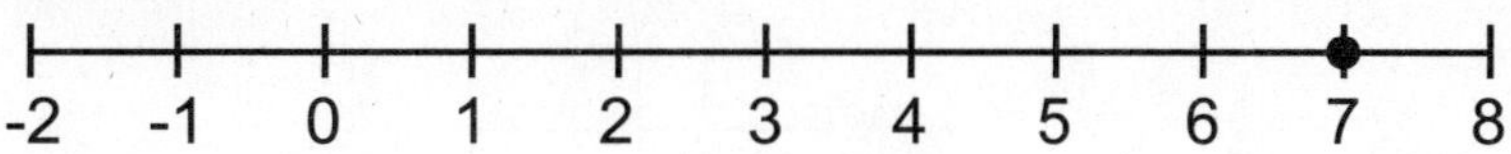

When you solve equations like $2x - 4 = 10$, you find one solution. From your work in Section 3.1, Exercise 3.2, you found that (6,6), (2,0) and $\left(-3, \frac{-15}{2}\right)$ were *all* solutions to $3x - 2y = 6$. In fact, a linear equation in two variables has an *infinite* number of solutions. The graph of a linear equation in two variables is a straight line, and a straight line is determined by two points.

To graph linear equations in two variables, do the following:

1. Find three ordered-pair solutions to the given equation. Completing either of the following tables is often convenient:

x	y	x	y
0		1	
	0	2	
1		3	

2. Plot the three points.

3. Draw a straight line through these points. If they are not in line, go back and check your arithmetic. Put arrows on both ends of the line to show that the line continues forever in both directions.

***Note*:** Although two points determine a line, you usually want to plot three as a check on your arithmetic.

Exercise 3.6

Graph:

a) $x - 2y = -6$

b) $x + y = 4$

c) $y = 2x$

Solution 3.6

a) $x - 2y = -6$

Let's complete the table:

x	y
0	
2	0

$x - 2y = -6$

If $x = 0$, $0 - 2y = -6$

x	y
0	3
2	0

$-2y = -6$

$y = 3$

If $y = 0$, $x - 2(0) = -6$

x	y
0	3
− 6	0
2	

$x = -6$

If $x = 2$, $(2) - 2y = -6$

x	y
0	3
− 6	0
2	4

$-2y = -8$

$y = 4$

The values in the table give the following ordered pairs as solutions to $x - 2y = -6$:

(0, 3)

(–6, 0)

(2, 4)

Now, plot your points and connect with a straight line.

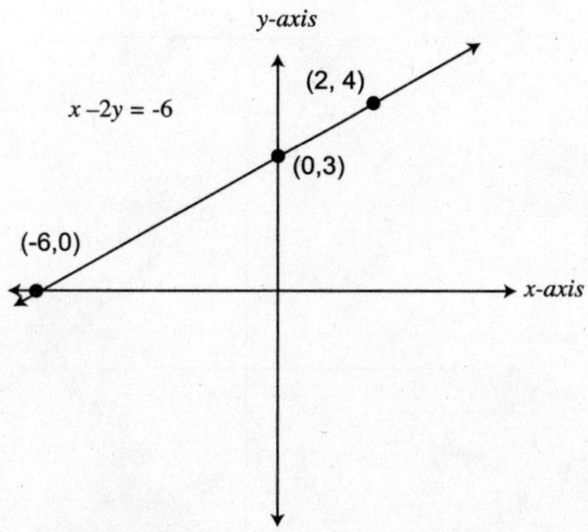

Notice that we used (2,) rather than (1,) to produce an integer solution. $\left(1, \frac{7}{2}\right)$ is, of course, also on the line.

b) $x + y = 4$

Let's complete the table:

x	y
1	
2	
3	

$x + y = 4$

If $x = 1$, $(1) + y = 4$

$y = 3$

x	y
1	3
2	2
3	1

If $x = 2$, $2 + y = 4$

$y = 2$

If $x = 3$, $3 + y = 4$

$y = 1$

The values in the table give the following ordered pairs as solutions to $x + y = 4$:
(1, 3)

(2, 2)

(3, 1)

Now, plot these points and connect with a straight line.

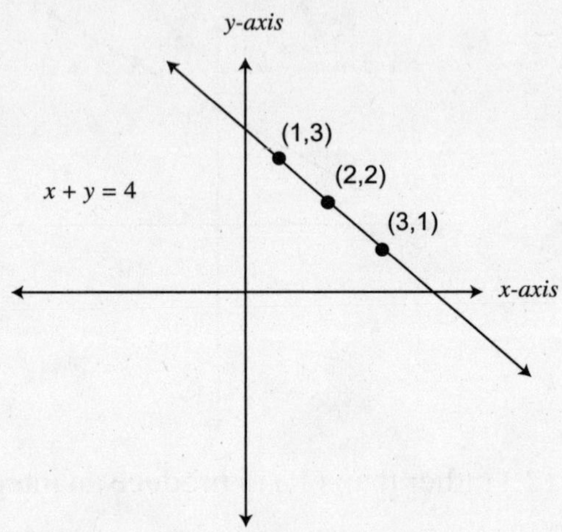

c) $y = 2x$

Let's complete the table:

x	y
1	
2	
3	

If $x = 1$, $y = 2(1)$

$y = 2$

If $x = 2$, $y = 2(2)$

$y = 4$

If $x = 3$, $y = 2(3)$

$y = 6$

The values in the table give the following ordered pairs as solutions to $y = 2x$:

(1, 2)

(2, 4)

(3, 6)

Now, plot these points and connect with a straight line.

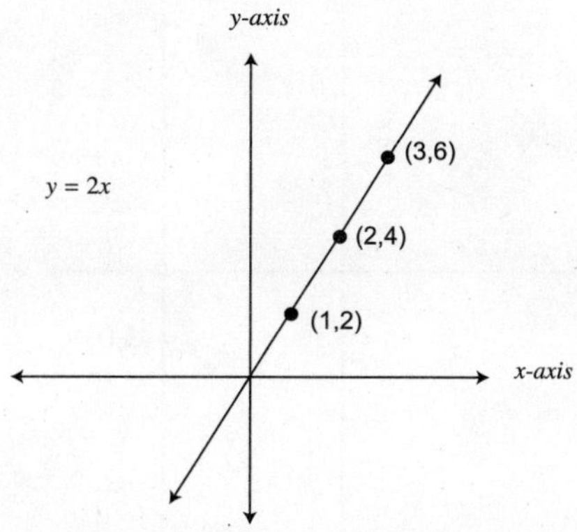

If you had used the table

x	y
0	
	0
1	

you would have found

x	y
0	0
0	0
1	2

and to find a third test point, you would have needed an extra x- or y-coordinate in your table.

Graphing Horizontal or Vertical Lines

The equation $x = 3$ has solutions (3,1), (3,3), (3, −2), and so on, and graphs as a vertical line. The equation, $y = -1$ graphs as a horizontal line. If you recognize these situations, you do not need a table of values.

To graph $x = a$, draw a vertical line through the point $(a, 0)$.
To graph $y = b$, draw a horizontal line through the point $(0, b)$.

Exercise 3.7

Graph:

a) $x = 3$

b) $y = -1$

Solution 3.7

a) To graph $x = 3$, locate the point (3, 0). Then draw a vertical line through (3, 0).

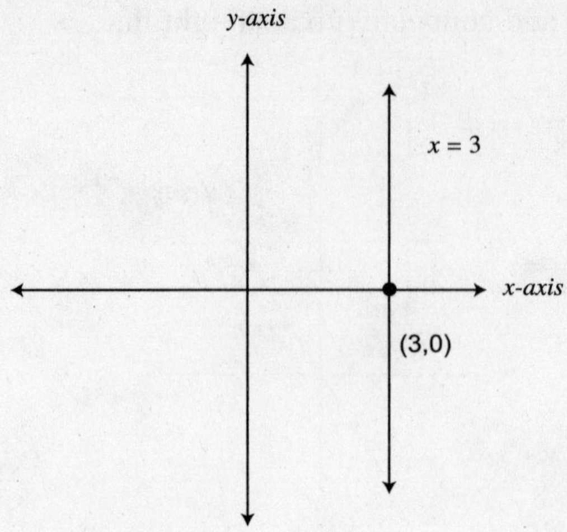

b) To graph $y = -1$, locate the point $(0, -1)$. Then draw a horizontal line through $(0, -1)$.

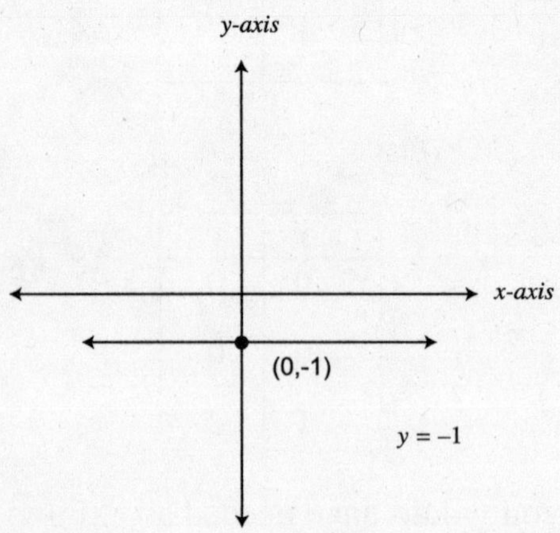

3.3 INTERCEPTS AND SLOPE

Two important features of a linear equation are the x- and y-intercepts and the slope. The following section explains how to find these features.

INTERCEPTS

The ***x*-intercept** of a graph is the x-coordinate where the graph crosses the x-axis. The ***y*-intercept** of a graph is the y-coordinate where the graph crosses the y-axis. To find the x-intercept, replace y with 0 in the equation. To find the y-intercept, replace x with 0. We generally use the variable b to represent the y-intercept. There is no standard variable used to represent the x-intercept. These intercepts give us a convenient way to graph a linear equation.

To graph a linear equation using the x- and y-intercepts:

1. Complete the table:

x	y	
0		This is the y-intercept
	0	This is the x-intercept

2. Connect the intercepts with a straight line.

In the preceding section, we used three points to graph linear equations. Although two points determine a line, we often use a third point as a check. You may want to use a third point to check the following graphs:

Exercise 3.8

Find the x- and y-intercepts for each linear equation and use them to draw the graph:

a) $2x - y = 2$

b) $y = \frac{3}{2}x + 3$

c) $y = 2x$

Solution 3.8

a) $2x - y = 2$

If $x = 0$, $2(0) - y = 2$ — Find the y-intercept, b.

$-y = 2$

$y = -2$ — $(0, -2)$

If $y = 0$, $2x - (0) = 2$ — Find the x-intercept.

$2x = 2$

$x = 1$ — $(1, 0)$

So the table becomes

x	y	
0	−2	The y-intercept
1	0	The x-intercept

Graph the intercepts and draw the line:

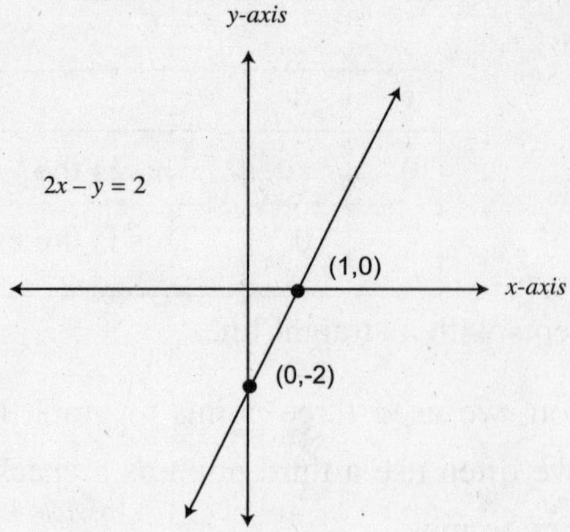

b) $y=\frac{3}{2}x+3$

If $x=0, y=\frac{3}{2}(0)+3$	Find the y-intercept, b.
$y=3$	(0, 3)
If $y=0, 0=\frac{3}{2}x+3$	Find the x-intercept.
$-3=\frac{3}{2}x$	
$\frac{2}{3}(-3)=\frac{2}{3}\left(\frac{3}{2}\right)x$	Multiply by $\frac{2}{3}$.
$-2=x$	(−2, 0)

So the table becomes

x	y	
0	–3	The y-intercept
–2	0	The x-intercept

Graph the intercepts and draw the line:

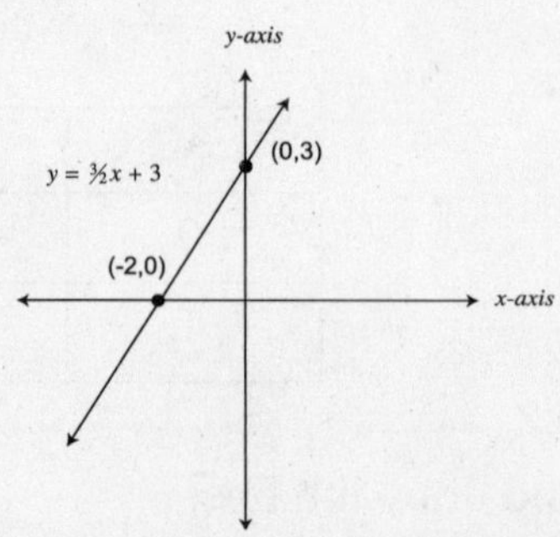

c) $y = 2x$

If $x = 0$, $y = 2(0)$	Find the y-intercept, b.
$y = 0$	(0, 0)
If $y = 0$, $0 = 2x$	Find the x-intercept.
$0 = x$	(0, 0)

Thus, (0, 0) is the x- and y-intercept. To graph the line, we will need at least one more point.

If $x = 1$, $y = 2(1)$	Substitute $x = 1$ into the original equation.
$y = 2$	(1, 2) is another point on the line.

So the table becomes

x	y	
0	0	The x- and y-intercept
1	2	

Graph the intercepts and draw the line:

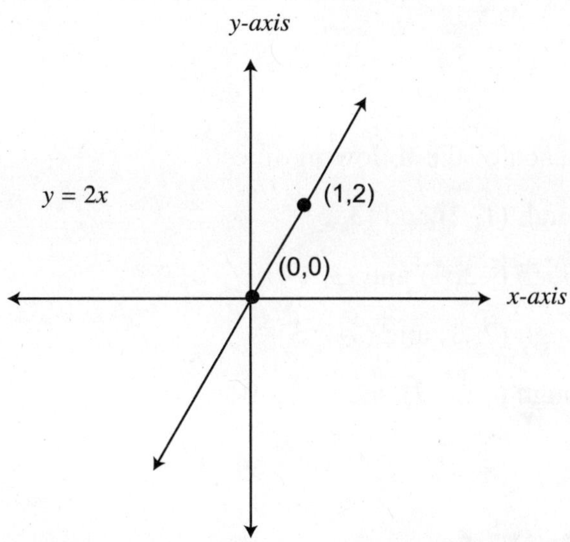

SLOPE

When we want to work with several points on a line, but not with specific coordinates, we use notation involving subscripts. Thus, two points on a line could be written as $P_1(x_1, y_1)$ and $P_2(x_2, y_2)$ and shown on a graph as follows:

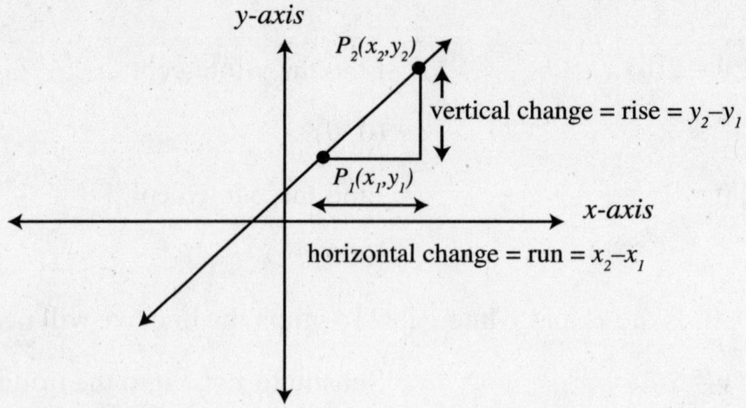

P_1 is read "*P* sub 1," x_1 is read "*x* sub 1", and y_2 is read " *y* sub 2," and so on; the 1 and 2 are called **subscripts.**

The **slope** of a line is a measure of the steepness of the line. It is calculated by the ratio $\frac{\text{vertical change}}{\text{horizontal change}}$.

The slope of a line containing points $P_1(x_1, y_1)$ and $P_2(x_2, y_2)$ is:

$$\text{Slope} = m = \frac{rise}{run} = \frac{y_2 - y_1}{x_2 - x_1} = \frac{y_1 - y_2}{x_1 - x_2}$$

Note that the letter *m* is the standard variable used to represent slope. Memorize the preceding formula.

Exercise 3.9

Find the slope of each of the following lines:

a) The line through (1, 2) and (3, 5)

b) The line through (–2, 1) and (3, –4)

c) The line through (2, 3) and (–1, 3)

d) The line through (–1, –2) and (–1, 4)

Solution 3.9

a) (x_1, y_1) (x_2, y_2) Label the points as (x_1, y_1) and (x_2, y_2).

(1, 2) (3, 5)

$m = \frac{y_2 - y_1}{x_2 - x_1}$ Write the formula for slope.

$= \frac{5-2}{3-1}$ Substitute $y_2 = 5$, $y_1 = 2$, and $x_2 = 3$, $x_1 = 1$.

$= \frac{3}{2}$ Simplify the numerator and denominator.

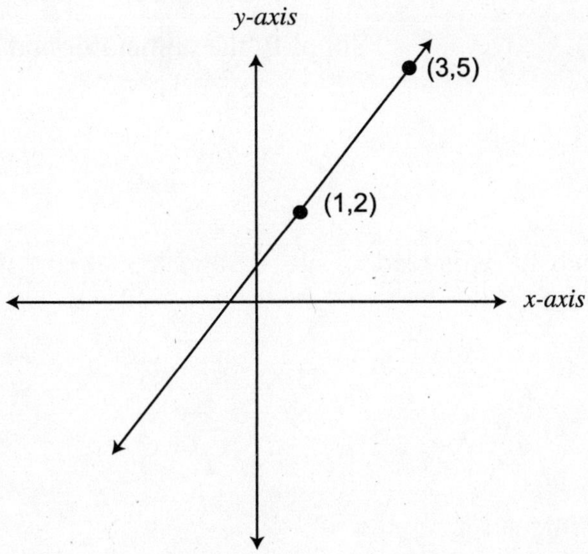

This line has a positive slope. When the slope of a line is positive, it *rises* from left to right.

If you had labeled the points as

(x_1, y_1) (x_2, y_2)

(3, 5) (1, 2)

your work would have looked like:

$m = \frac{y_2 - y_1}{x_2 - x_1}$ Write the formula for slope.

$= \frac{2-5}{1-3}$ Substitute $y_2 = 2$, $y_1 = 5$, and $x_2 = 1$, $x_1 = 3$.

$= \frac{-3}{-2} = \frac{3}{2}$ Simplify the numerator and denominator.

Notice that the slope equals $\frac{3}{2}$ regardless of which point you choose as P_1 and which you choose as P_2.

b) (x_1, y_1) and (x_2, y_2). Label the points as (x_1, y_1) and (x_2, y_2).

(–2, 1) (3, –4)

$m = \dfrac{y_2 - y_1}{x_2 - x_1}$ Write the formula for slope.

$= \dfrac{-4-1}{3-(-2)}$ Substitute $y_2 = -4$, $y_1 = 1$, and $x_2 = 3$, $x_1 = -2$.

$= \dfrac{-5}{5}$ Simplify the numerator and denominator.

$= -1$

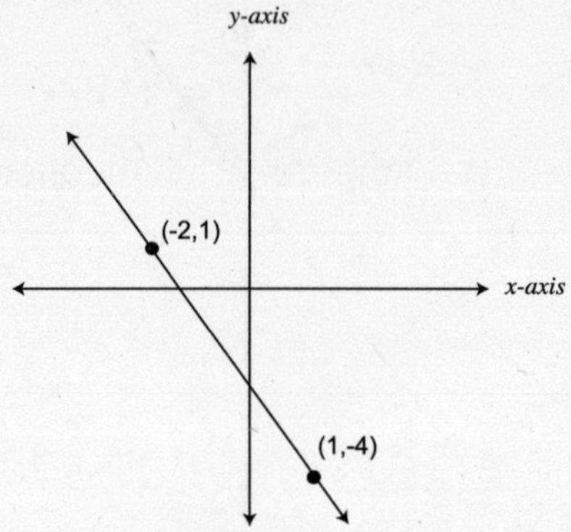

This line has a negative slope. When the slope of a line is negative, it *falls* from left to right.

c) (x_1, y_1) and (x_2, y_2). Label the points as (x_1, y_1) and (x_2, y_2).

(2, 3) (−1, 3)

$m = \dfrac{y_2 - y_1}{x_2 - x_1}$ Write the formula for slope.

$= \dfrac{3-3}{-1-2}$ Substitute $y_2 = 3$, $y_1 = 3$, and $x_2 = -1, x_1 = 2$

$= \dfrac{0}{-3} = 0$ Simplify the numerator and denominator.

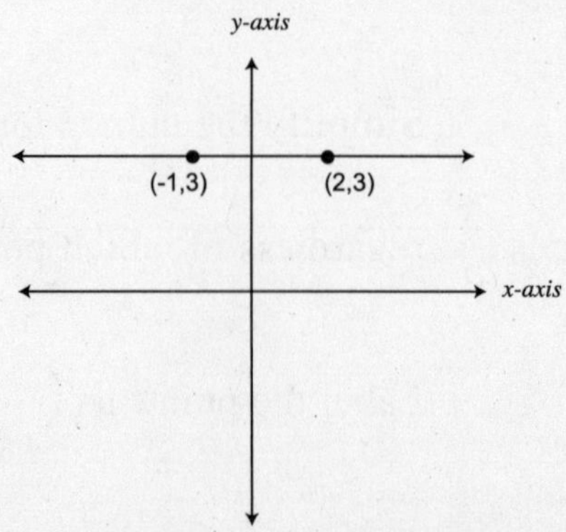

This line has 0 slope. When the slope is 0, the graph is a *horizontal* line.

d) (x_1, y_1) and (x_2, y_2) Label the points as (x_1, y_1) and (x_2, y_2).

$(-1, -2)\ (-1, 4)$

$m = \dfrac{y_2 - y_1}{x_2 - x_1}$ Write the formula for slope.

$= \dfrac{4-(-2)}{-1-(-1)}$ Substitute $y_2 = 4$, $y_1 = -2$, and $x_2 = -1, x_1 = -1$.

$= \dfrac{6}{0}$ Simplify the numerator and denominator; this answer is undefined.

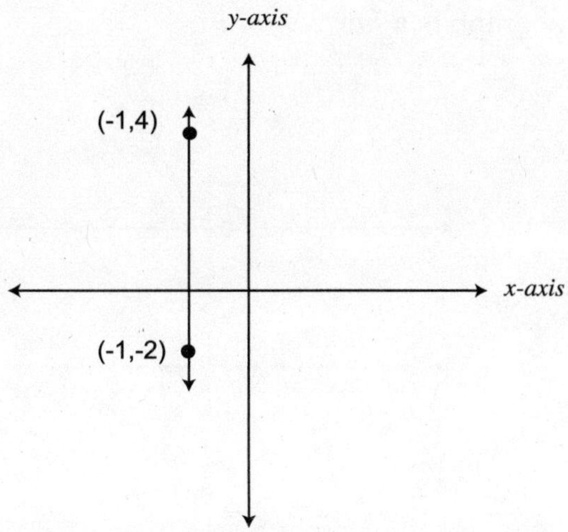

This line has undefined slope. When the slope is undefined, the graph is a *vertical* line.

Here are some slope concepts to remember:

* **Positive slope:** Rises from left to right

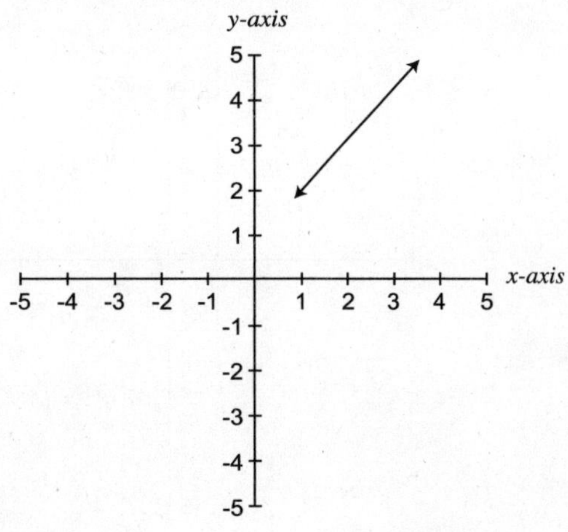

* **Negative slope:** Falls from left to right

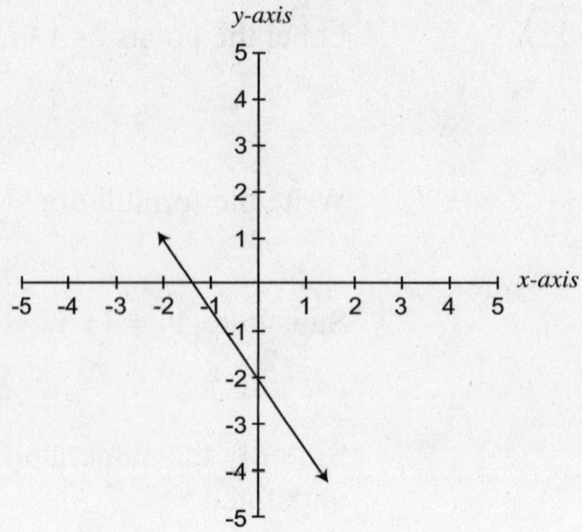

When the slope = 0, the graph is a *horizontal* line.

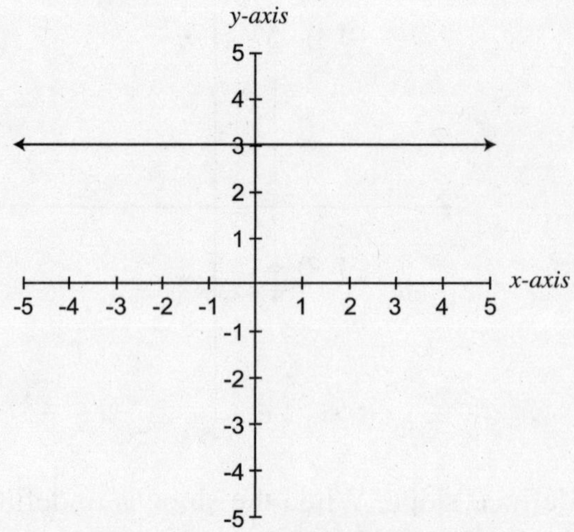

When the slope is undefined, the graph is a *vertical* line.

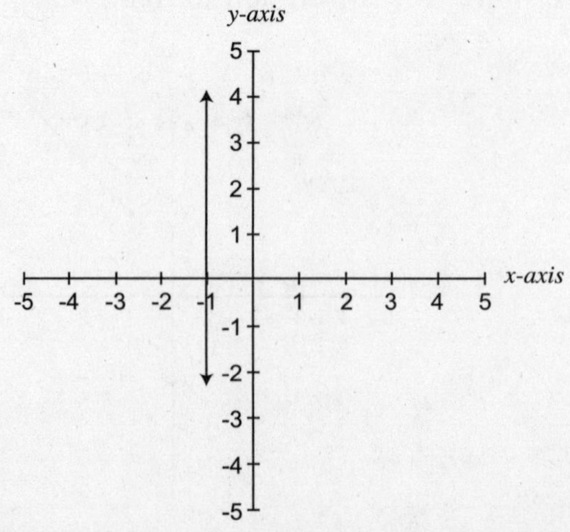

GRAPHING USING THE SLOPE AND A POINT ON THE LINE

If you know the slope of a line and any point on the line, you can graph the line using the following steps:

To graph the line through the point (x_1, y_1) with slope $m = \frac{a}{b}$:

1. Put a point on (x_1, y_1) and use this as your starting point.

2. Use $m = \frac{rise}{run} = \frac{a}{b}$ and rise (or go up) a units from your starting point. Then move to the right b units and put a second point there.

3. Draw a line through the two points.

***Note*:** If the given slope is an integer, write it as a fraction using 1 for the denominator. If the given slope is negative, carry the negative in the numerator and *fall* (go down) a units instead of rising a units.

Exercise 3.10

Graph the lines with the given slopes and containing the given points:

a) $m = \frac{3}{2}$, (–1, –2)

b) $m = 2$, (0, 1)

c) $m = -\frac{1}{3}$, (–1, 2)

Solution 3.10

a) Start at (–1, –2). $m = \frac{3}{2} = \frac{rise}{run}$, rise (go up) 3 units, and run 2 units to the right.

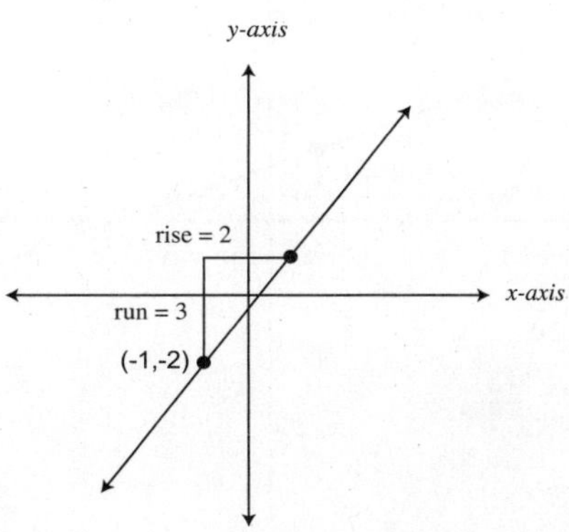

b) Start at (0, 1).

$m = 2$ — This is the given slope.

$= \frac{2}{1}$ — Rewrite 2 as the fraction $\frac{2}{1}$.

$\frac{2}{1} = \frac{rise}{run}$ — Rise (go up) 2 units, and run 1 unit to the right.

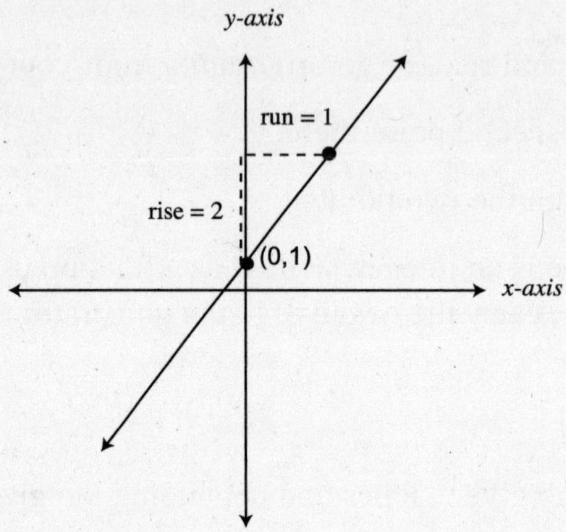

c) Start at (–1, 2).

$m = -\frac{1}{3}$ — This is the given slope.

$= \frac{-1}{3}$ — Rewrite the negative fraction with the negative in the numerator.

$\frac{-1}{3} = \frac{rise}{run}$ — Fall (go down) 1 unit, and run 3 units to the right.

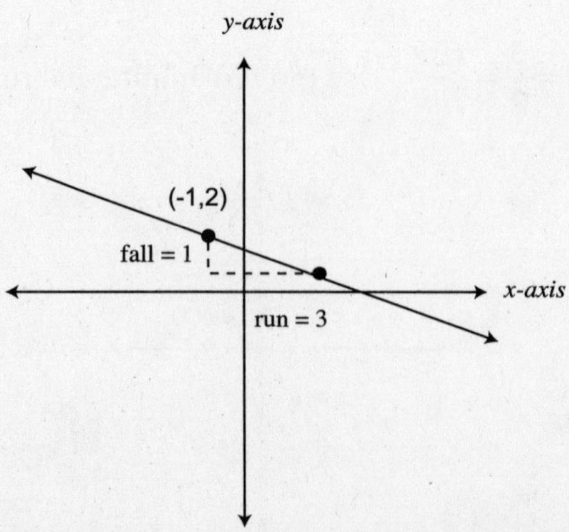

Note in (c) that $-\frac{1}{3} = \frac{-1}{3} = \frac{1}{-3}$, so you can *rise* 1 unit, run to the *left* 3 units, and draw the line.

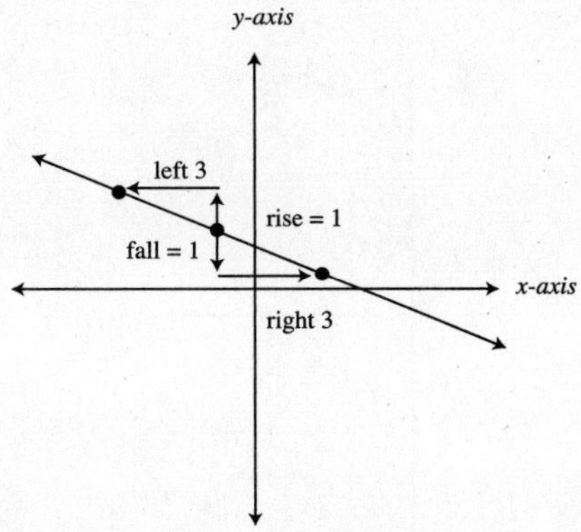

If you are given the y-intercept, start on the y-axis, use the slope to find a second point, and then graph the line.

Exercise 3.11

Graph the line with the given slope and y-intercept b:

(a) $m = \frac{1}{2}, b = 1$

(b) $m = -\frac{3}{2}, b = 2$

(c) $m = 4$, $b = -3$

(d) $m = -2$, $b = 0$

Solution 3.11

(a) Start at (0, 1). $m = \frac{1}{2} = \frac{rise}{run}$, so rise (go up) 1 unit, and run 2 units to the right.

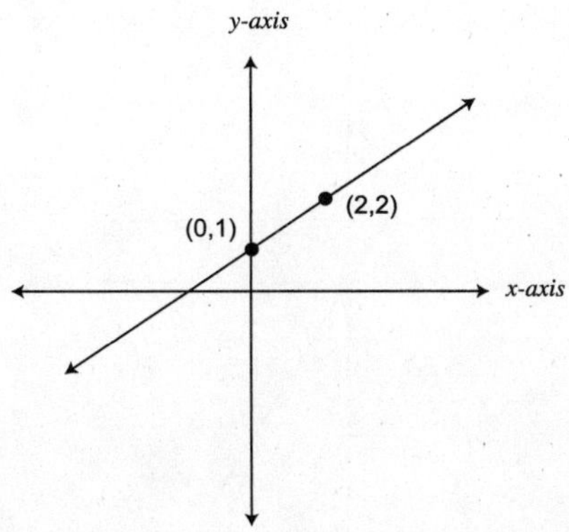

(b) Start at (0, 2). $m = \dfrac{-3}{2} = \dfrac{rise}{run}$ fall (go down) 3 units, and run 2 units to the right.

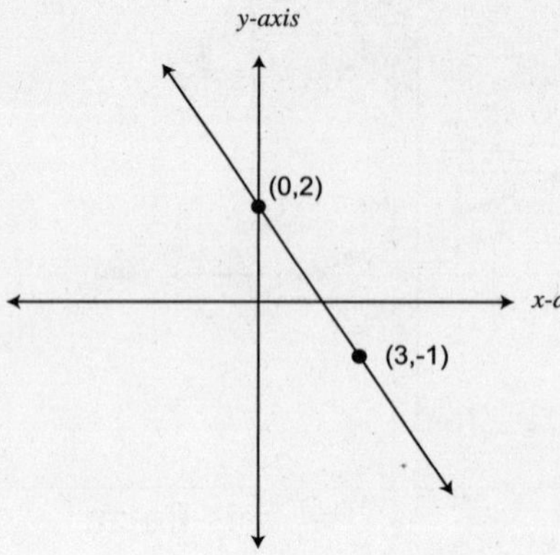

(c) Start at (0, –3). $m = 4 = \dfrac{4}{1} = \dfrac{rise}{run}$ rise (go up) 4 units, and run 1 unit to the right.

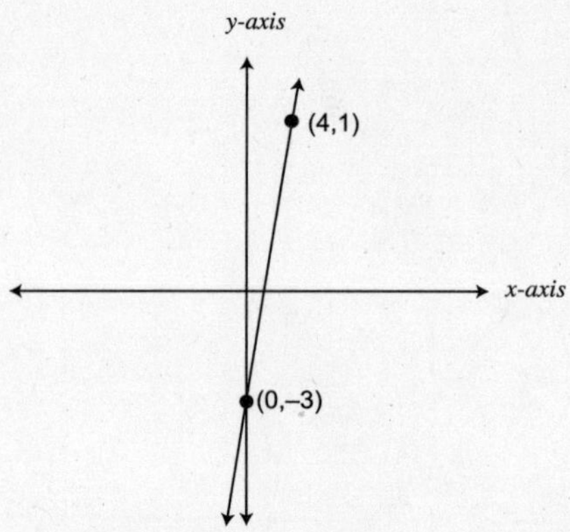

(d) Start at (0, 0). $m = -2 = \dfrac{-2}{1} = \dfrac{rise}{run}$, fall (go down) 2 units, and run 1 unit to the right.

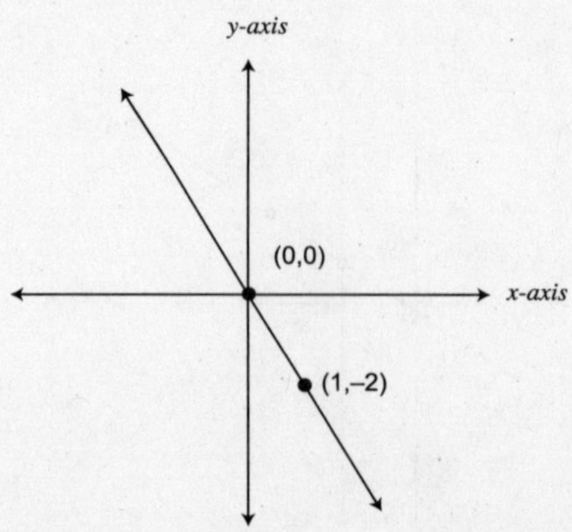

3.4 FINDING THE EQUATION OF A LINE

In the previous section you were given the equation of the line and were asked to graph the line. The reverse situation, finding the equation of the line, is presented next.

EQUATIONS OF LINES

In this section, you will find the equation of a line, given various pieces of information. The table below helps you decide when to use each form of a linear equation. You must memorize these forms.

Given This Data	Use This Form	Name of the Form
Slope m and y-intercept b	$y = mx + b$	Slope-intercept
Point (x_1, y_1) and slope m	$y - y_1 = m(x - x_1)$	Point-slope
Two points (x_1, y_1) and (x_2, y_2)	First find $m = \frac{y_2 - y_1}{x_2 - x_1}$ then use $y - y_1 = m(x - x_1)$	Slope of a line Point-slope
Horizontal line through (x_1, y_1)	$y = y_1$	
Vertical line through (x_1, y_1)	$x = x_1$	

Exercise 3.12

Write an equation of the line using the given information:

a) $m = 2$, y-intercept = 1

b) $m = \frac{3}{2}$, containing the point (–1, –2)

c) Containing the points (–1, –2) and (2, 1)

d) A horizontal line passing through (3, 4)

e) A vertical line passing through (3, 4)

Solution 3.12

a) $m = 2,\ b = 1$ — This is the given slope and y-intercept.

$y = mx + b$ — This is in slope-intercept form.

$y = 2x + 1$ — Substitute $m = 2$ and $b = 1$ into the slope-intercept form of a line.

b) $m = \frac{3}{2}, (-1,-2)$ — This is the given slope and a point.

$y - y_1 = m(x - x_1)$ — This is in point-slope form.

$y - (-2) = \frac{3}{2}(x - (-1))$ — Substitute $m = \frac{3}{2},\ x_1 = -1,\ y_1 = -2$.

$y + 2 = \frac{3}{2}(x + 1)$ — This is the equation of the line.

c) $(x_1, y_1) \quad (x_2, y_2)$ — These are the two given points.

$(-1, 2) \quad (2,\ 1)$

$m = \frac{y_2 - y_1}{x_2 - x_1} = \frac{1-2}{2-(-1)}$ — First find the slope.

$m = -\frac{1}{3}$

$y - y_1 = m(x - x_1)$ — Now use the point-slope form of a line.

$y - 2 = -\frac{1}{3}(x - (-1))$ — Substitute $y_1 = 2,\ m = -\frac{1}{3}$, and $x_1 = -1$.

$y - 2 = -\frac{1}{3}(x + 1)$ — This is the equation of the line.

If you had labeled your points

$(x_1, y_1) \quad (x_2, y_2)$

$(2, 1) \quad (-1, 2)$

your work would look like:

$m = \frac{y_2 - y_1}{x_2 - x_1} = \frac{2-1}{(-1)-2}$ — First find the slope.

$m = \frac{1}{-3} = -\frac{1}{3}$

$y - y_1 = m(x - x_1)$ Now use the point-slope form of a line.

$y - 1 = -\frac{1}{3}(x - 2)$ Substitute $y_1 = 1$, $m = -\frac{1}{3}$, and $x_1 = 2$.

$y - 1 = -\frac{1}{3}(x - 2)$ This is the equation of the line.

Both of these answers are correct. In the next exercise, we will show that $y - 2 = -\frac{1}{3}(x + 1)$ and $y - 1 = -\frac{1}{3}(x - 2)$ are equivalent equations. Thus, you may label the points either way and get the correct answer.

d) Horizontal line through (3, 4)

$y = 4$ This is the equation of the line.

e) Vertical line through (3, 4)

$x = 3$ This is the equation of the line.

FORMS OF LINEAR EQUATIONS

All the answers given in Exercise 3.12 are linear equations in two variables. You may need to write your answers in slope-intercept or standard form. After you have found the equation, use the addition and multiplication properties of equality to rewrite the equation in the desired form.

$ax + by = c$ Standard form

$y = mx + b$ Slope-intercept form

Exercise 3.13

Rewrite each equation in the required form:

a) $y = 2x + 1$ in standard form

b) $y + 2 = \frac{3}{2}(x + 1)$ in slope-intercept form

c) $y - 2 = -\frac{1}{3}(x + 1)$ in slope-intercept form

d) $y - 1 = -\frac{1}{3}(x - 2)$ in slope-intercept form

Solution 3.13

a) $y = 2x + 1$ — We need $ax + by = c$.

$y - 2x = 2x + 1 - 2x$ — Get x and y on one side.

or

$2x - y = -1$ — We multiply both sides by –1 so that the first number in the equation is positive.

b) $y + 2 = \frac{3}{2}(x+1)$ — We need $y = mx + b$.

$y + 2 = \frac{3}{2}x + \frac{3}{2}$ — Distribute $\frac{3}{2}$.

$y = \frac{3}{2}x + \frac{3}{2} - 2$ — Subtract 2 from both sides: $\frac{3}{2} - \frac{2}{1} = \frac{3}{2} - \frac{4}{2} = -\frac{1}{2}$.

$y = \frac{3}{2}x - \frac{1}{2}$ — $y = mx + b$ form.

c) $y - 2 = -\frac{1}{3}(x+1)$ — We need $y = mx + b$.

$3(y-2) = 3\left[-\frac{1}{3}(x+1)\right]$ — Multiply by the LCD, 3, to clear fractions.

$3y - 6 = -1(x + 1)$ — Multiply by 3.

$3y - 6 = -x - 1$ — Distribute.

$3y - 6 + \mathbf{6} = -x - 1 + \mathbf{6}$ — Add 6 to both sides.

$3y = -x + 5$ — Combine similar terms.

$\frac{1}{3}(3y) = \frac{1}{3}(-x+5)$ — Multiply both sides by $\frac{1}{3}$.

$y = -\frac{1}{3}x + \frac{5}{3}$

d) $y - 1 = -\frac{1}{3}(x-2)$ — We need $y = mx + b$.

$3(y-1) = 3\left[-\frac{1}{3}(x-2)\right]$ — Multiply by the LCD, 3, to clear fractions.

$3y - 3 = -1(x - 2)$	Multiply by 3.
$3y - 3 = -1x + 2$	Distribute the -1.
$3y - 3 + \mathbf{3} = -x + 2 + \mathbf{3}$	Add 3 to both sides.
$3y = -x + 5$	Combine similar terms.
$\frac{1}{3}(3y) = \frac{1}{3}(-x+5)$	Multiply both sides by $\frac{1}{3}$.
$y = -\frac{1}{3}x + \frac{5}{3}$	

Compare this answer to the answer for (c). Note that when the two equations $y-2=-\frac{1}{3}(x+1)$ and $y-1=-\frac{1}{3}(x-2)$ are put in slope-intercept form, they are the same equation.

USING SLOPE-INTERCEPT TO GRAPH LINEAR EQUATIONS

We can now use the slope-intercept form of a line to graph lines, because this form gives you a point $(0, b)$ and the slope, m. Remember to solve for y first, if necessary.

Exercise 3.14

Graph each equation, using the slope and y-intercept:

a) $-3x + y = 2$

b) $2x - 5y = 20$

Solution 3.14

a) $-3x + y = 2$	This is the original equation.
$-3x + y + \mathbf{3x} = \mathbf{3x} + 2$	Add $3x$ to both sides.
$y = 3x + 2$	Slope-intercept form: $y = mx + b$.

So $m = 3 = \frac{3}{1}$ and $b = 2$.

Start at $(0, 2)$; $m = \frac{3}{1} = \frac{rise}{run}$.

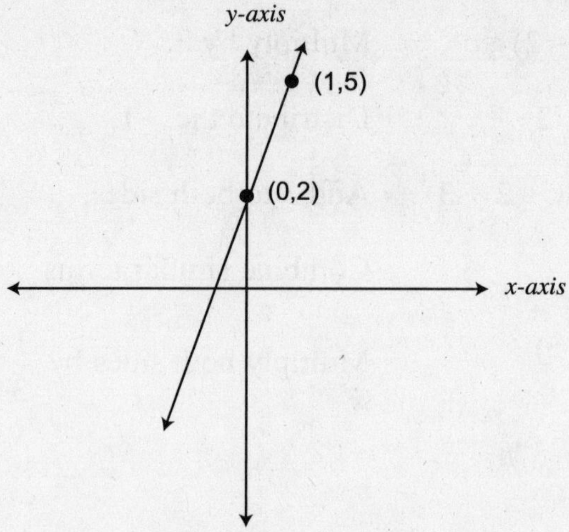

b) $2x - 5y = 20$ This is the original equation.

$2x - 5y - \mathbf{2x} = \mathbf{-2x} + 20$ Subtract $2x$ from both sides.

$-5y = -2x + 20$ Combine similar terms.

$-\frac{1}{5}(-5y) = -\frac{1}{5}(-2x+20)$ Multiply both sides by $-\frac{1}{5}$.

$y = \frac{2}{5}x - 4$ Distribute $-\frac{1}{5}$.

Slope-intercept form: $y = mx + b$.

So $m = \frac{2}{5}$ and $b = -4$.

Start at (0, –4), $m = \frac{2}{5} = \frac{rise}{run}$.

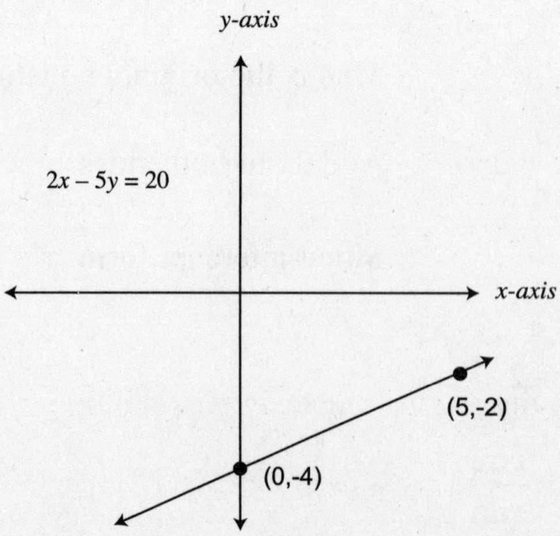

You can also use slope to determine whether lines are horizontal or vertical. If a line contains the point (x_1, y_1) and has $m = 0$, it must be a horizontal line with the equation $y = y_1$. If a line contains the point (x_1, y_1) and has undefined slope, it must be a vertical line with the equation $x = x_1$.

Exercise 3.15

Find the equation of a line with the given characteristics:

a) The line contains the point (–2, 5) and $m = 0$.

b) The line contains the point (–4, –1) and has undefined slope.

Solution 3.15

a) Because $m = 0$, this must be a horizontal line with equation $y = 5$.
b) Because the slope is undefined, this must be a vertical line with equation $x = -4$.

PARALLEL AND PERPENDICULAR LINES

The last use of slope in this section involves parallel and perpendicular lines.

* Parallel lines have equal slopes.
* Perpendicular lines have slopes that are negative reciprocals of each other (a and $-\frac{1}{a}$, for $a \neq 0$).

Exercise 3.16

Determine whether each pair of lines is parallel, perpendicular, or neither:

a) $y = 2x - 4;\ -4x + 2y = 12$

b) $x - 2y = 10;\ y = -2x + 3$

c) $y = 3x + 1;\ 3x + y = -4$

d) $x = 4;\ x = 2$

e) $x = -1;\ y = 3$

Solution 3.16

First find the slope of each line by writing each equation in $y = mx + b$ form.

a) Equation 1: $y = 2x - 4$ This is already in $y = mx + b$ form.
$m = 2$

Equation 2: $-4x + 2y = 12$ Get y alone on one side.

$-4x + 2y + \mathbf{4x} = \mathbf{4x} + 12$ Add $4x$ to both sides.

$2y = 4x + 12$ Combine similar terms.

$\frac{1}{2}(2y) = \frac{1}{2}(4x + 12)$ Multiply both sides by $\frac{1}{2}$.

$y = 2x + 6$

$m = 2$

The lines have equal slopes of 2, so the lines are parallel.

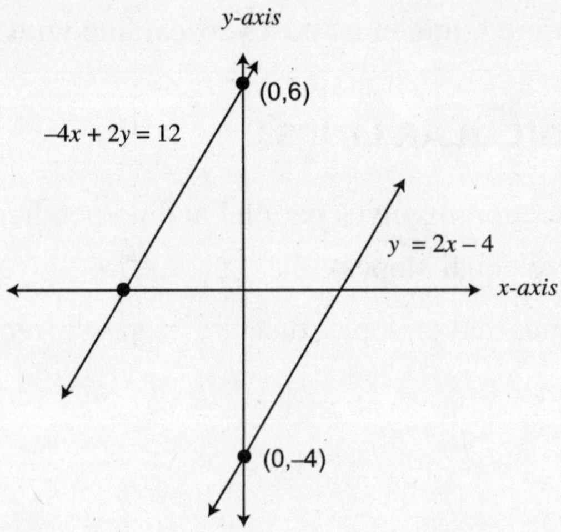

b) Equation 1:

$x - 2y = 10$ Get y alone on one side.

$x - 2y - \mathbf{x} = -\mathbf{x} + 10$ Subtract x from both sides.

$-2y = -x + 10$ Combine similar terms.

$-\frac{1}{2}(-2y) = -\frac{1}{2}(-x + 10)$ Multiply both sides by $-\frac{1}{2}$.

$y = \frac{1}{2}x - 5$

$m = \frac{1}{2}$.

Equation 2:

$y = -2x + 3$ Already in $y = mx + b$ form.

$m = -2$

The slopes of the lines are negative reciprocals, so the lines are perpendicular.

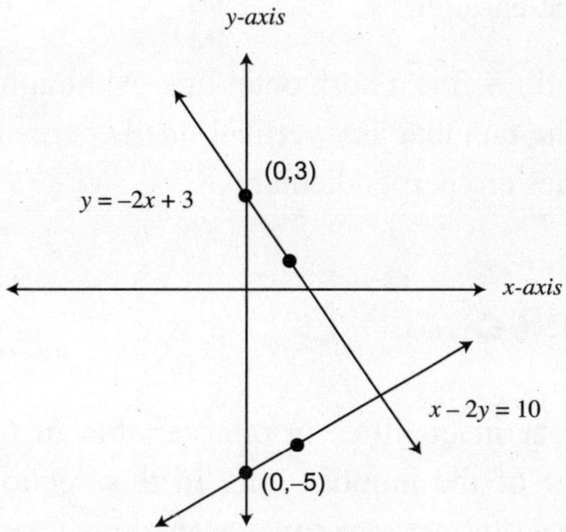

c) Equation 1:

$y = 3x + 1$ Already in $y = mx + b$ form.

$m = 3$

Equation 2:

$3x + y = -4$ Get y alone on one side.

$3x + y \mathbf{- 3x} = \mathbf{-3x} - 4$ Subtract $3x$ from both sides.

$y = -3x - 4$

$m = -3$

The slopes are *not* equal, so the lines are *not* parallel. The slopes are *not* negative reciprocals, so the lines are *not* perpendicular.

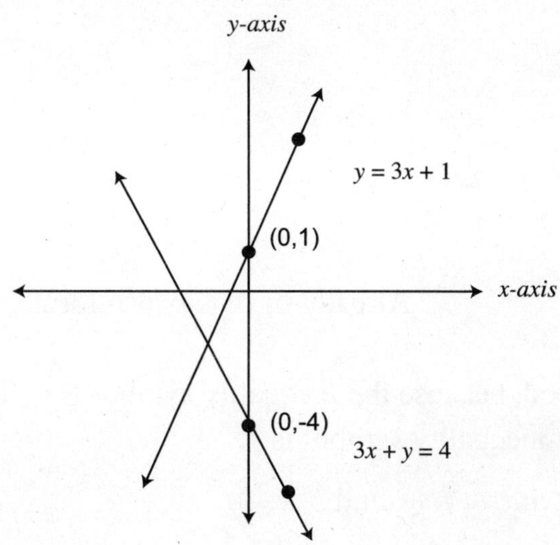

d) $x = 4$ is a vertical line, and $x = 2$ is a vertical line. Vertical lines have undefined slope, but all vertical lines will be parallel.

e) $x = -1$ is a vertical line, and $y = 3$ is a horizontal line. Although you can't apply the negative reciprocal concept, the fact that any vertical line is perpendicular to any horizontal line tells you that these lines are perpendicular.

3.4 GRAPHING LINEAR INEQUALITIES

Recall that you graphed linear inequalities in one variable in Chapter 2, and that these graphs involved shading a part of the number line. In this section, you will graph linear inequalities in two variables, and these graphs involve shading a part of the rectangular coordinate system.

GRAPHING LINEAR INEQUALITIES IN TWO VARIABLES

To graph a linear inequality in two variables:

1. Rewrite the inequality in the form $y \square mx + b$, where the box contains $<$, $>$, $\leq$, or $\geq$.
2. Graph a solid line at $y \square mx + b$ if the box contains $\leq$ or $\geq$. Graph a dotted line at $y \square mx + b$ if the box contains $<$ or $>$.
3. Shade above the line if the box contains $>$ or $\geq$. Shade below the line if the box contains $<$ or $\leq$.

Exercise 3.17

Graph each inequality:

a) $y < \frac{1}{2}x - 3$

b) $y \geq -2x + 1$

c) $2x - y < 4$

d) $y \leq 3x$

Solution 3.17

a) $y < \frac{1}{2}x - 3$ Already in $y = mx + b$ form.

The line will be dotted, because the inequality symbol is $<$. The shading will be below the line, because the inequality symbol is $<$.

Graph a dotted line with $m = \frac{1}{2}$ and $b = -3$.

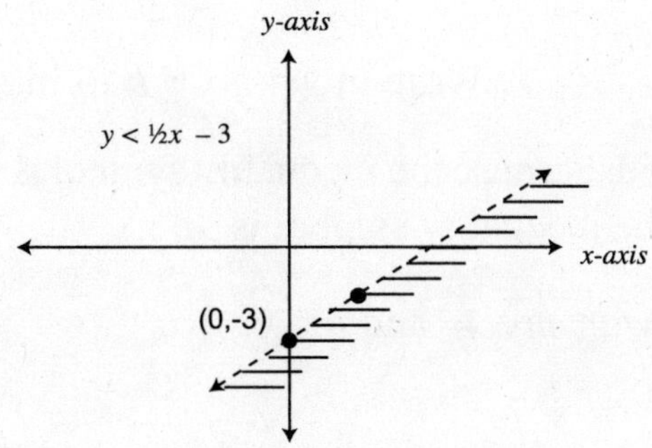

b) $y \geq -2x + 1$ Already in $y = mx + b$ form.

The line will be solid, because the inequality symbol is ≥. The shading will be above the line, because the inequality symbol is ≥.

Graph a solid line with $m = -\frac{2}{1}$ and $b = 1$.

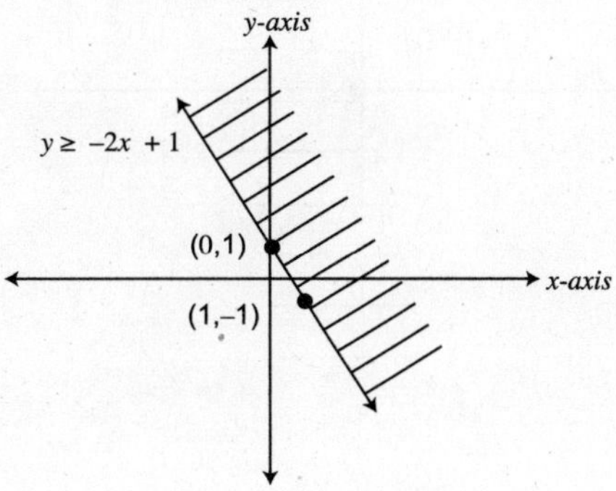

c) $2x - y < 4$ Rewrite in $y = mx + b$ form.

$2x - y - \mathbf{2x} < -\mathbf{2x} + 4$ Subtract $2x$ from both sides.

$-y < -2x + 4$ Combine similar terms.

$-1(-y) > -1(-2x + 4)$ Reverse the inequality symbol when multiplying by a negative.

$y > 2x - 4$

The line will be dotted, because the inequality symbol is >. The shading will be above the line, because the inequality symbol is >.

Graph a dotted line with $m = \frac{2}{1}$ and $b = -4$.

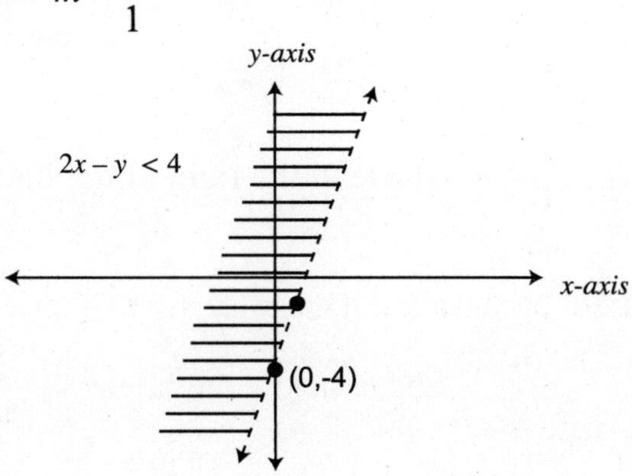

d) $y \leq 3x$
$y \leq 3x + 0$ Write in $y = mx + b$ form.

The line will be solid, because the inequality symbol is ≤. The shading will be below the line, because the inequality symbol is ≤.

Graph a solid line with $m = \frac{3}{1}$ and $b = 0$.

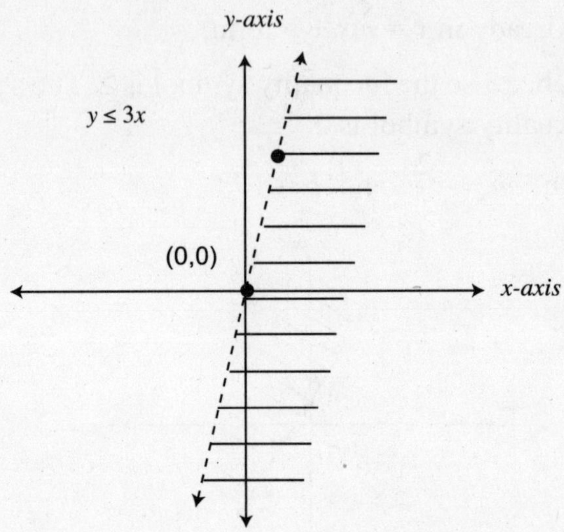

GRAPHING HORIZONTAL AND VERTICAL LINEAR INEQUALITIES

To graph horizontal and vertical linear inequalities:

1. Rewrite the inequality in the form $x \square a$ or $y \square b$ where $\square$ contains <, >, ≤, or ≥.

2. Graph a solid line at $x \square a$ or $y \square b$ if the box contains ≤ or ≥. Graph a dotted line at $x \square a$ or $y \square b$ if the box contains < or >.

3. Shade as follows:

$\left\{\begin{matrix} x < a \\ \text{or} \\ x \leq a \end{matrix}\right\}$ Shade to the left of the line.

$\left\{\begin{matrix} x > a \\ \text{or} \\ x \geq a \end{matrix}\right\}$ Shade to the right of the line.

$\left\{\begin{matrix} y < b \\ \text{or} \\ y \leq b \end{matrix}\right\}$ Shade below the line.

$$\begin{Bmatrix} y > b \\ \text{or} \\ y \geq b \end{Bmatrix}$$ Shade above the line.

Exercise 3.18

Graph each inequality:

a) $x < -2$

b) $x - 3 \geq 1$

c) $y < 1$

d) $2y + 6 \geq 0$

Solution 3.18

a) $x < -2$ Dotted line at $x = -2$; shade to the left of the line.

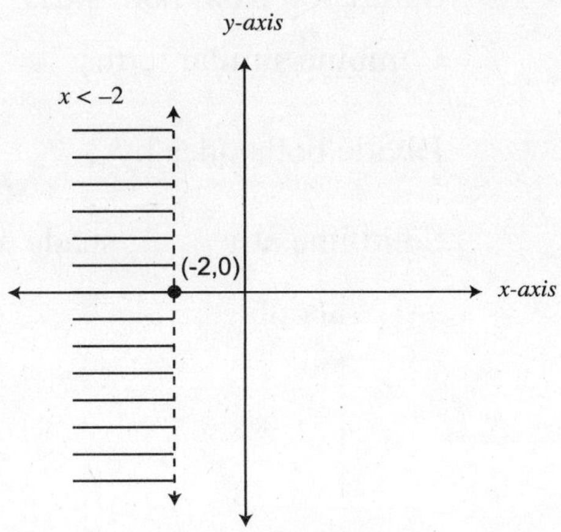

b) $x - 3 \geq 1$

$x - 3 + \mathbf{3} \geq 1 + \mathbf{3}$ Add 3 to both sides.

$x \geq 4$ Solid line at $x = 4$; shade to the right of the line.

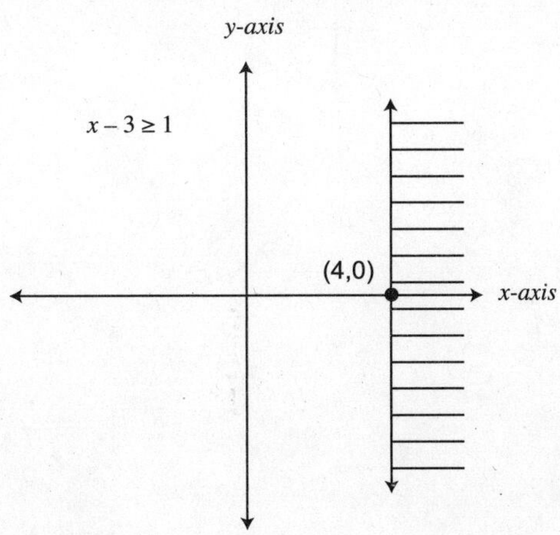

c) $y < 1$ Dotted line at $y = 1$; shade below the line.

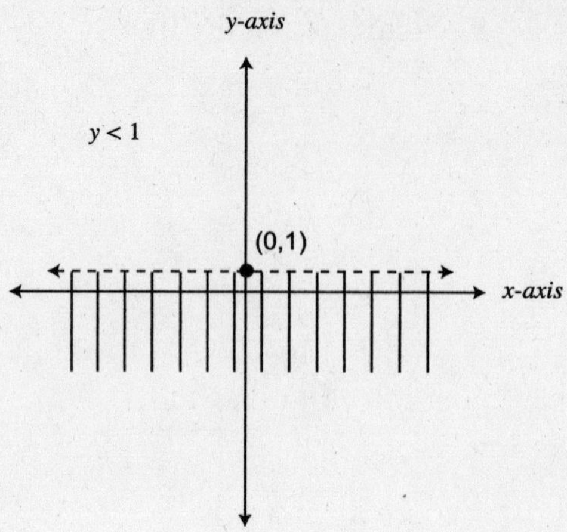

d) $2y + 6 \geq 0$	This is the original inequality.
$2y + 6 - \mathbf{6} \geq 0 - \mathbf{6}$	Subtract 6 from both sides.
$2y \geq -6$	Combine similar terms.
$\frac{2y}{2} \geq \frac{-6}{2}$	Divide both sides by 2.
$y \geq -3$	Solid line at $y = -3$; shade above the line.

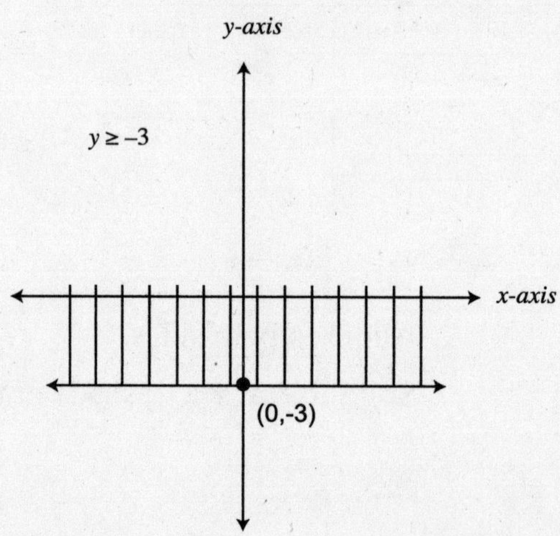

TEST YOURSELF

1) Determine whether the given ordered pair is a solution of the given equations:

   (a) $4x - 3y = 10$; $(1, -2)$

   (b) $3x - 2y = 17$; $(3, 4)$

   (c) $4x - 10y = -4; \left(\frac{1}{2}, \frac{1}{5}\right)$

   (d) $2x + 3y = -4$; $(-5, 2)$

2) Complete the given ordered pairs for the equations:

   (a) $3x - 5y = -25$; $(x, 2)$

   (b) $6x + 4y = -2; (-\frac{5}{3}, y)$

   (c) $12x - 3y = -4$; $(x, 4)$

   (d) $2x - y = 9$; $(3, y)$

3) Graph the ordered pairs:

   (a) $(2, 5)$

   (b) $(-1, 3)$

   (c) $(-2, -4)$

   (d) $(4, -1)$

   (e) $(3, 0)$

   (f) $(0, 2)$

4) Give the coordinates of each labeled point:

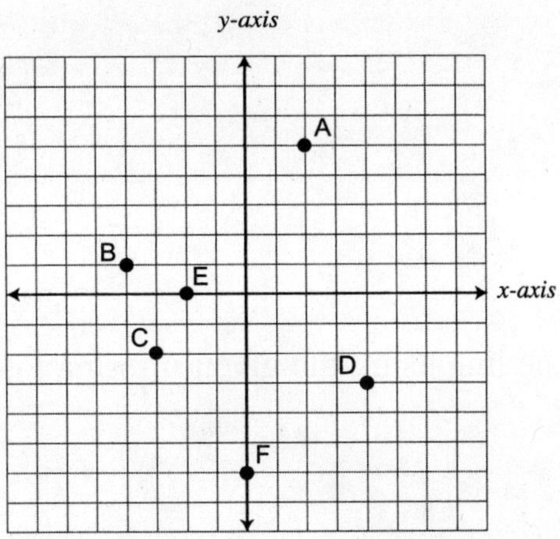

5) Graph:

   (a) $2x + y = 4$

(b) $x = 8$

(c) $3x - y = -3$

(d) $y = 4x$

6) Graph:

(a) $4x - 2y = 6$

(b) $x - y = -6$

(c) $3x - y = 2$

(d) $2x = y$

7) Find the slope of each of the following lines:

(a) The line through (5, 1) and (2, –1)

(b) The line through (2, –3) and (3, 12)

(c) The line through (–1, 4) and (0, 4)

(d) The line through (2, 6) and (2, –3)

8) Graph the line with the given slope and containing the given point:

(a) $m = \frac{1}{2}; (2, 4)$

(b) $m = 3$; (0, 1)

(c) $m = -\frac{2}{5}; (2, -3)$

(d) $m = 0$; (–1, –7)

9) Graph the line with the given slope and y-intercept b:

(a) $m = \frac{2}{3}; b = -2$

(b) $m = -\frac{5}{3}; b = 3$

(c) $m = -2$; $b = -1$

(d) $m = 0$; $b = 3$

10) Write an equation of the line using the given information. Write each answer in standard form, if possible:

(a) $Q(-2, 6); m = -\frac{2}{5}$

(b) $P(0, 0); m = \frac{2}{3}$

(c) Through the points A (1, 3) and B(5, –3)

(d) Through A (–4, –1) and parallel to the graph of $3x + 4y = 12$

(e) A horizontal line through (4, –2)

11) Determine whether the pair of lines is parallel, perpendicular, or neither:

(a) $2x - 4y = 6;\ 2x + y = 8$

(b) $y = -\dfrac{2}{5}$; $5x + 2y = 1$

(c) $y = -3x + 4;\ 6x + 2y = 10$

(d) $x = 4;\ y = 5$

12) Graph each inequality:

(a) $2x - y \le 4$

(b) $3x + 4y < 8$

(c) $x + 3y \le 6$

(d) $2x - y < 4$

13) Graph each inequality:

(a) $x > 2$

(b) $y \le 3$

(c) $x > -4$

(d) $y < 0$

TEST YOURSELF ANSWERS

1) (a) Yes

   (b) No

   (c) No

   (d) Yes

2) (a) $x = -5$

   (b) $y = 2$

   (c) $x = \dfrac{2}{3}$

   (d) $y = -3$

3)

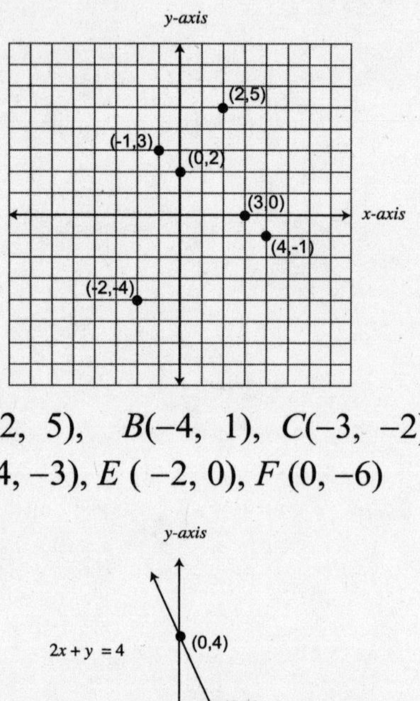

4) $A(2,\ 5)$, $B(-4,\ 1)$, $C(-3,\ -2)$, $D(4, -3)$, $E\ (-2, 0)$, $F\ (0, -6)$

5) (a)

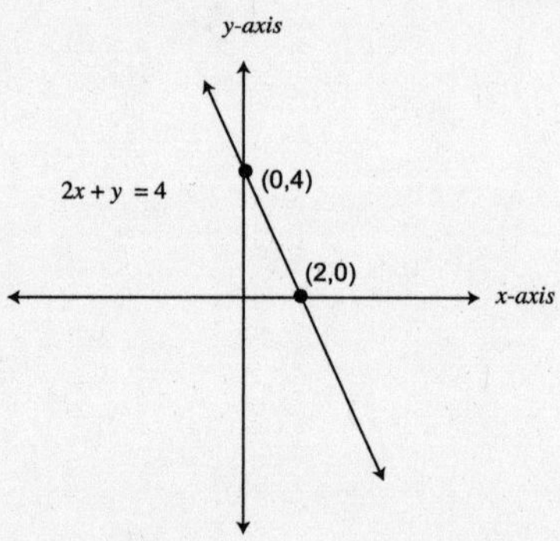

(b)

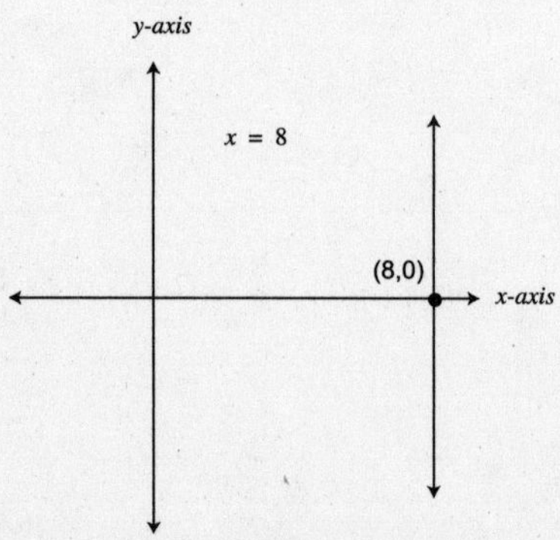

(c)

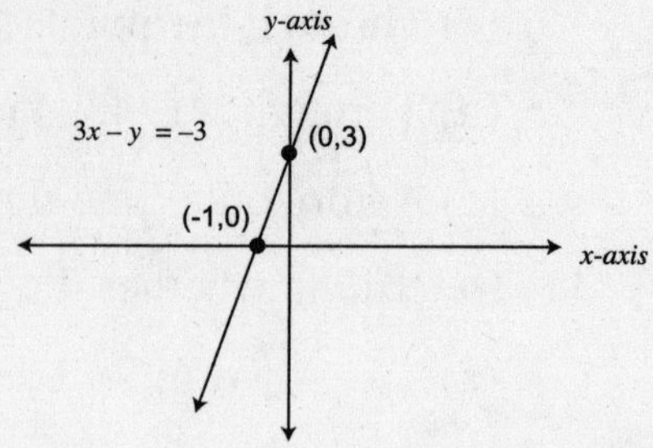

(d)

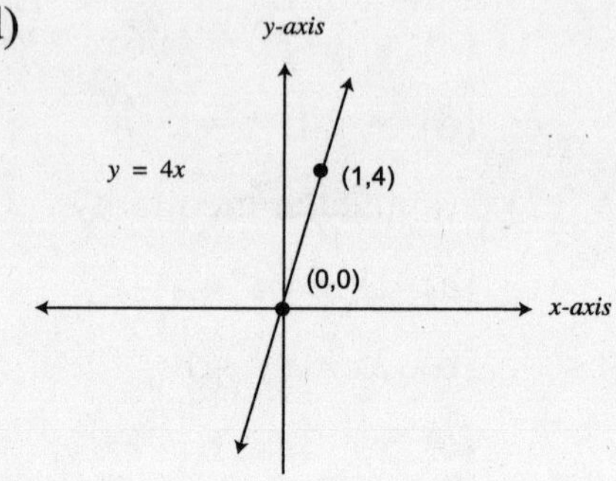

6) (a)

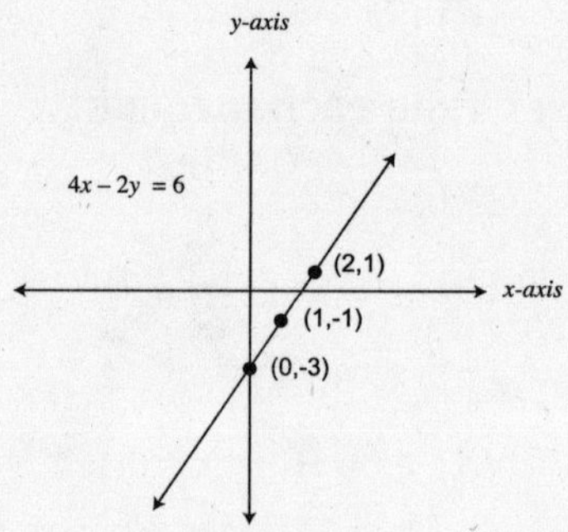

(b)

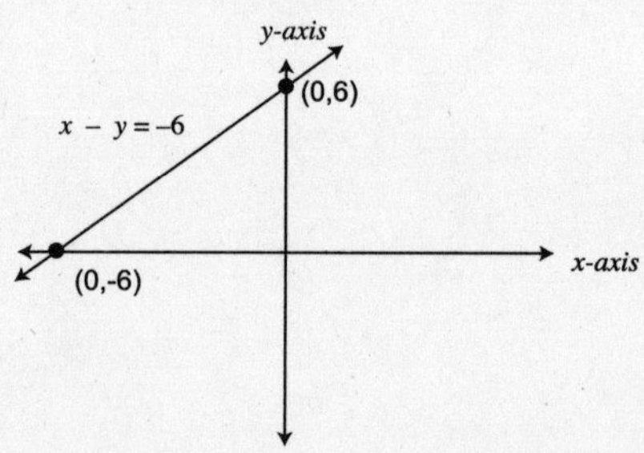

(c)

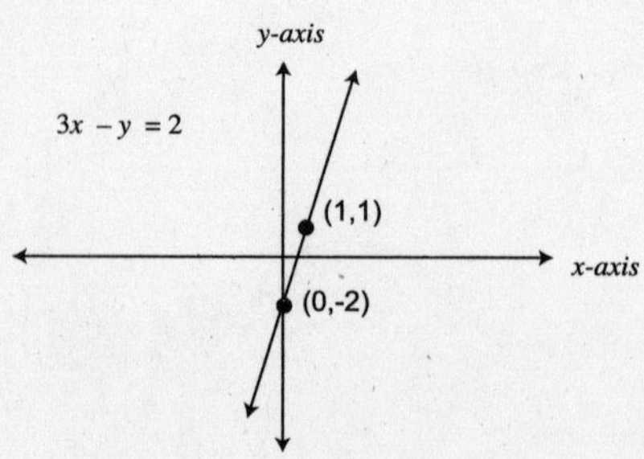

(d)

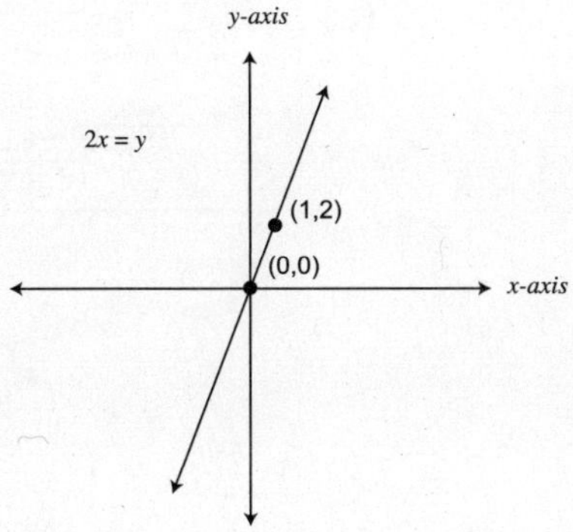

7) (a) $\frac{2}{3}$

(b) 15

(c) 0

(d) Undefined

8) (a)

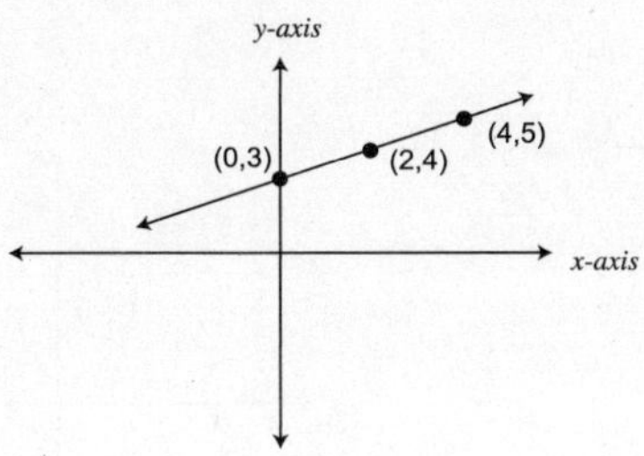

(b)

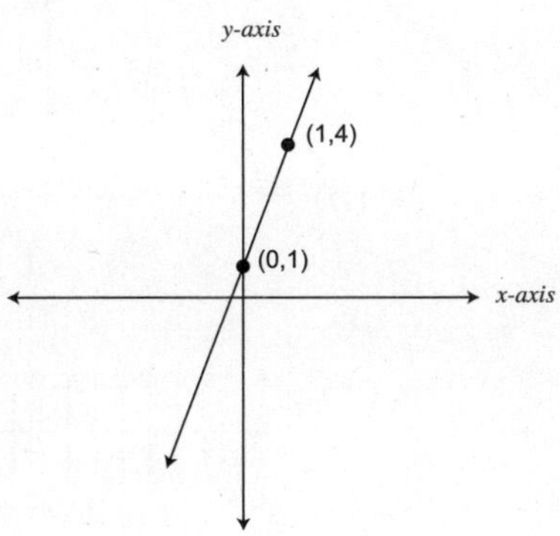

(c)

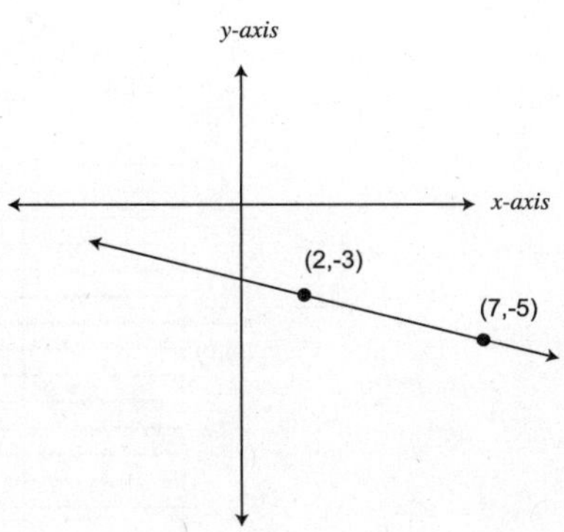

(d)

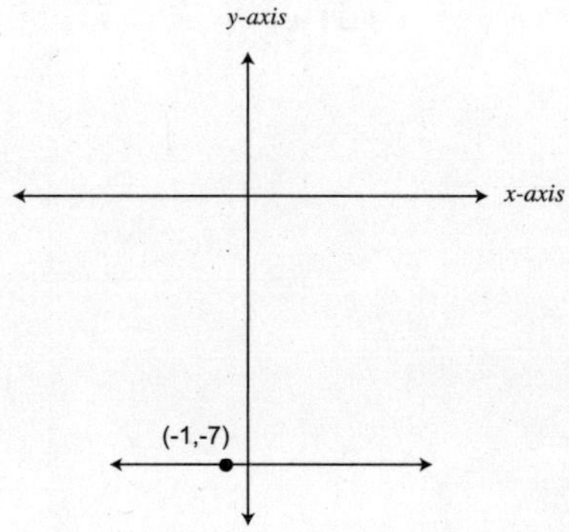

9) (a)

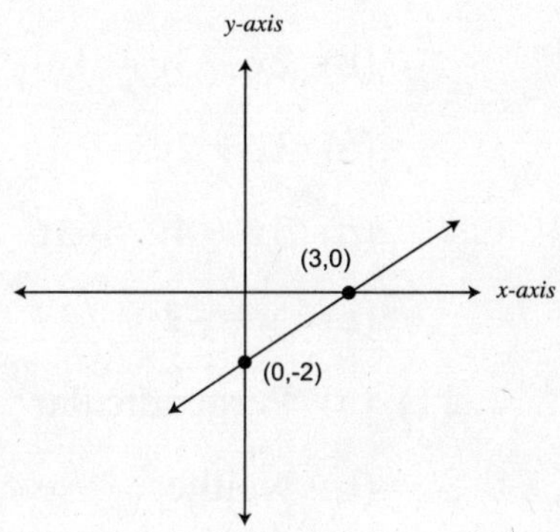

(b)

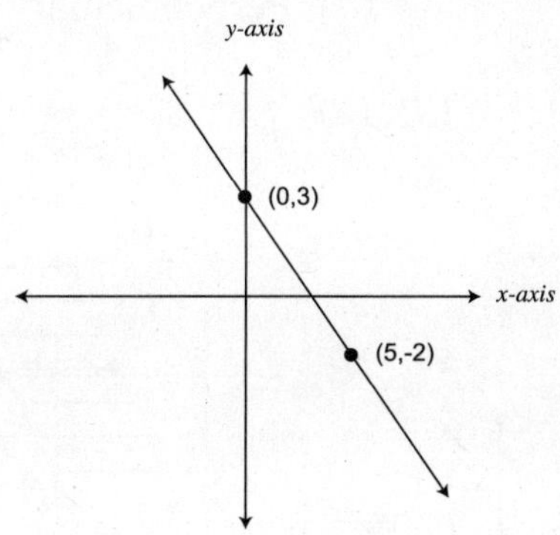

(c)

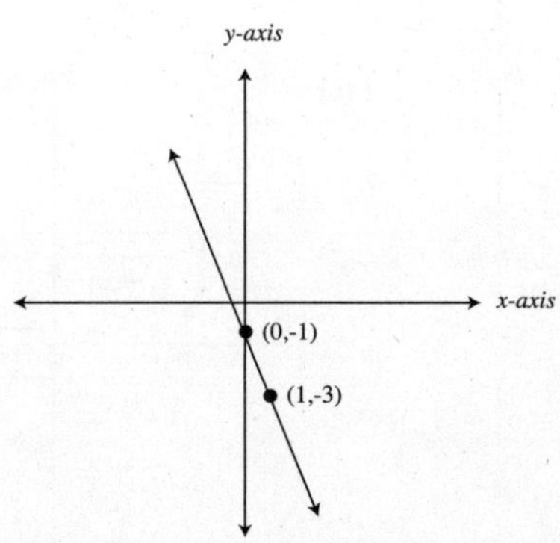

(d)

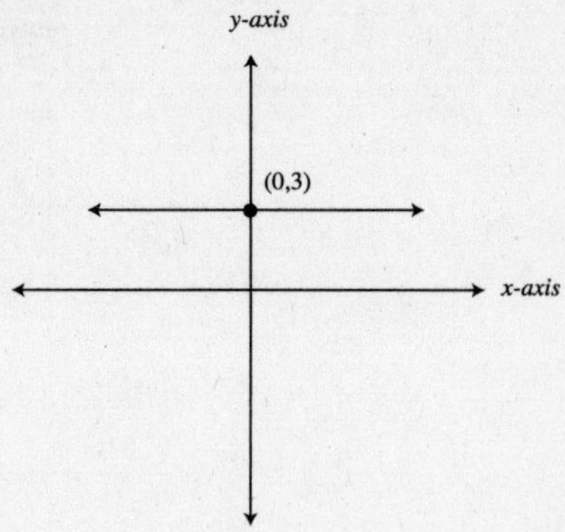

10) (a) $2x + 5y = 26$

(b) $2x - 3y = 0$

(c) $3x + 2y = 9$

(d) $3x + 4y = -16$

(e) $y = -2$

11) (a) Perpendicular

(b) Neither

(c) Parallel

(d) Perpendicular

12) (a)

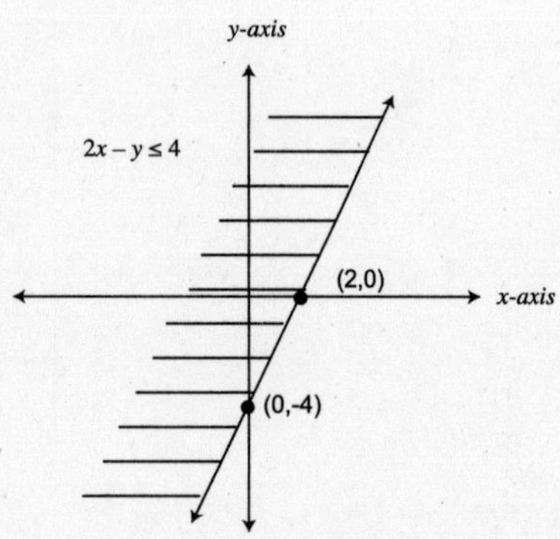

(b)

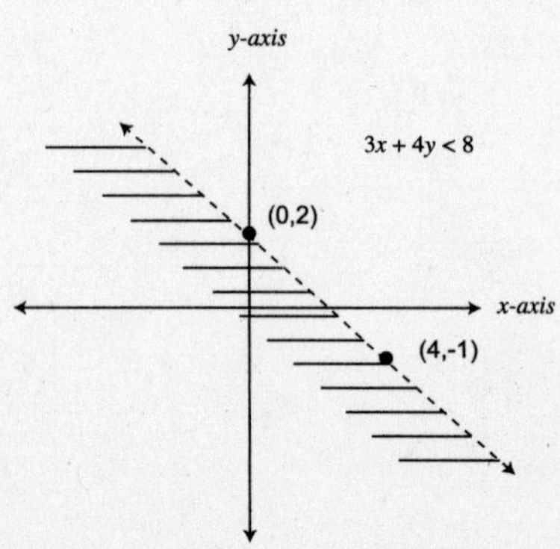

(c)

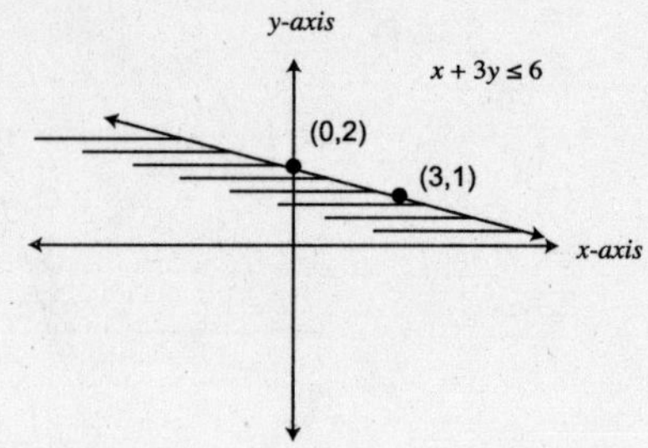

(d)

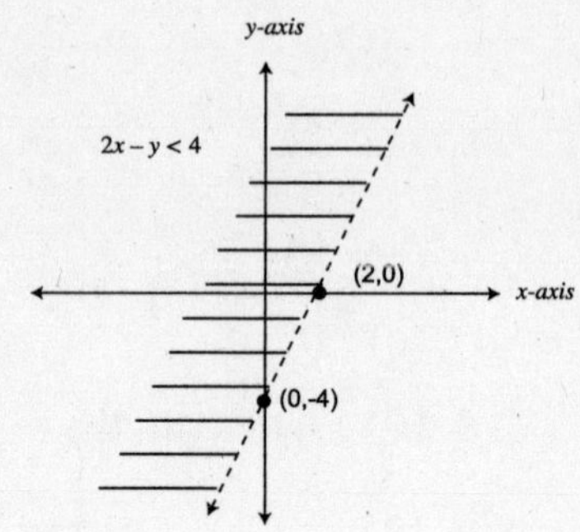

13) (a)

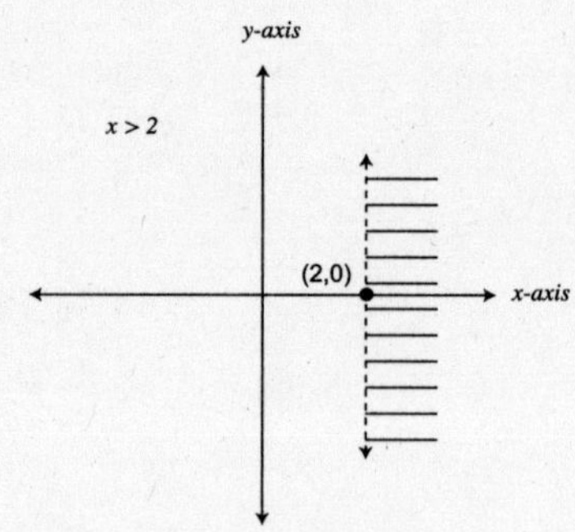

(b)

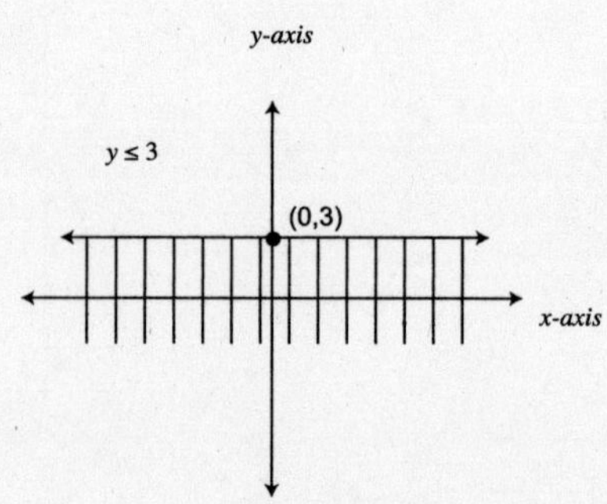

(c)

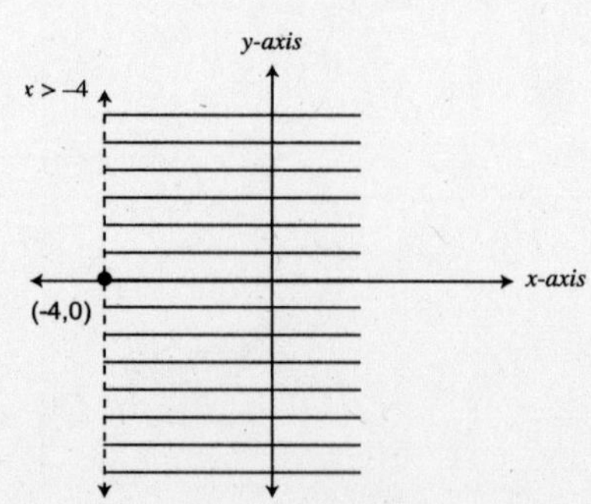

(d)

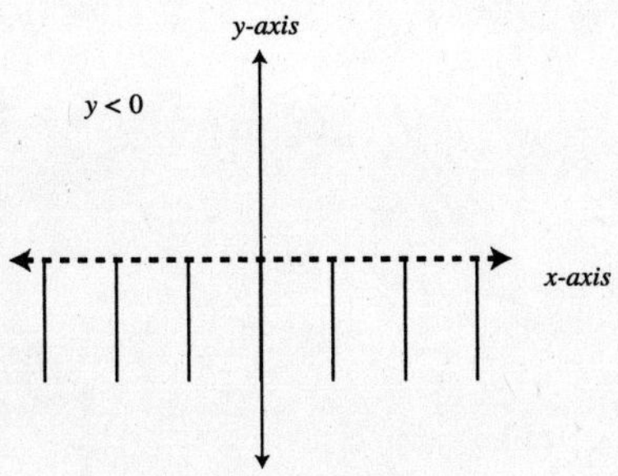

CHAPTER 4

Systems of Linear Equations

While many word problems can be solved by writing and solving one linear equation, other applications require two or more equations to describe the relationship of the data. This chapter describes several techniques for solving two equations in two unknowns.

4.1 SOLVING SYSTEMS OF LINEAR EQUATIONS BY GRAPHING

A visual approach to solving systems of linear equations requires graphing each equation and finding the point of intersection of two lines, if it exists. That point of intersection must be a solution to each equation.

DETERMINING WHETHER AN ORDERED PAIR IS A SOLUTION TO A SYSTEM OF LINEAR EQUATIONS

A system of linear equations in two variables consists of two linear equations in two unknowns, such as the following:

$3x + 2y = 7$

$x - 2y = -11$

Solving a system of linear equations in two variables means finding all the ordered pairs that are solutions to *both* equations.

Exercise 4.1

a) Is $(-1, 5)$ a solution to the following system?

$3x + 2y = 7$

$x - 2y = -11$

b) Is (2, 3) a solution to the following system?

$4x - 2y = 2$

$2x + y = 5$

c) Is (0, 6) a solution to the following system?

$5x + 2y = 12$

$2x - 3y = 18$

Solution 4.1

a) $3x + 2y = 7$

$3(-1) + 2(5) = 7$	Substitute $x = -1$, $y = 5$.
$-3 + 10 = 7$	True
$x - 2y = -11$	
$(-1) - 2(5) = -11$	Substitute $x = -1$, $y = 5$.
$-1 - 10 = -11$	True

Because (–1, 5) is a solution to both equations, (–1, 5) is a solution to the system.

b) $4x - 2y = 2$

$4(2) - 2(3) = 2$	Substitute $x = 2$, $y = 3$.
$8 - 6 = 2$	True
$2x + y = 5$	
$2(2) + 3 = 5$	Substitute $x = 2$, $y = 3$.
$4 + 3 = 5$	False

Because (2, 3) is not a solution to both equations, (2, 3) is not a solution to the system.

c) $5x + 2y = 12$

$5(0) + 2(6) = 12$	Substitute $x = 0$, $y = 6$.
$0 + 12 = 12$	True
$2x - 3y = 18$	
$2(0) - 3(6) = 18$	Substitute $x = 0$, $y = 6$.
$0 - 18 = 18$	
$-18 = 18$	False

Because (0, 6) is not a solution to both equations, (0, 6) is not a solution to the system.

GRAPHING SYSTEMS OF LINEAR EQUATIONS

When two lines are graphed on the same set of axes, there are three possible types of solutions. Consider the following graphs of two lines:

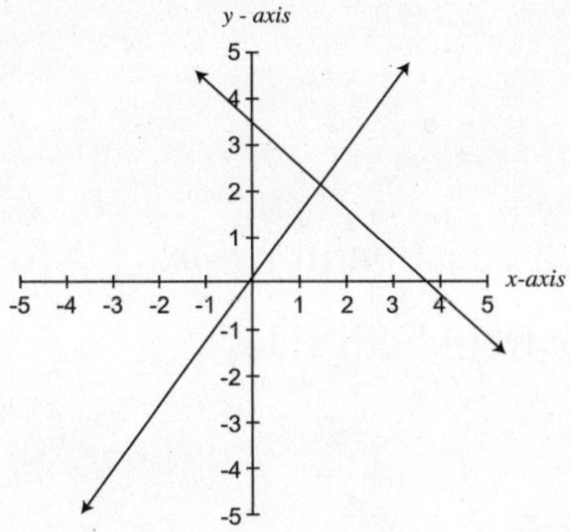

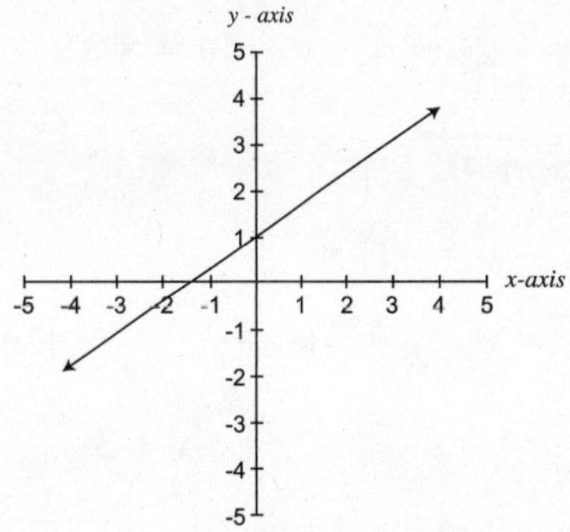

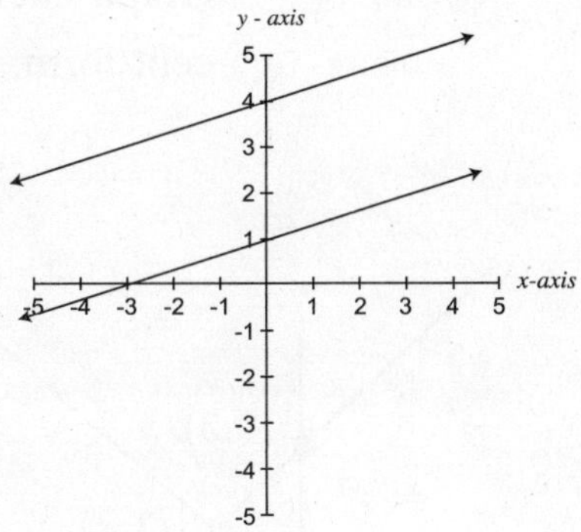

One Solution	No Solutions	Infinite Number of Solutions
The answer will be an ordered pair.	The answer will be ∅, the empty set.	The answer will be every point on the line.
Consistent system	Inconsistent system	Consistent system
Independent equations	Independent equations	Dependent equations

To find a solution to a linear system by graphing:

1. Write each equation in $y = mx + b$ form.
2. Graph both equations on the same set of x- and y-axes.
3. Look for the intersection point.
4. Check by substituting the ordered pair in each equation.

You must use a sharp pencil, graph paper, and a ruler to help you draw an accurate graph. And even with the best tools, your graph may be slightly off.

Exercise 4.2

Solve each system by graphing:

a) $x - y = -1;\ x + y = 3$

b) $y = -\frac{3}{2}x$; $x + 2y = 4$

c) $3x - y = 2$; $3x - y = -1$

d) $x + 2y = 4$; $2x + 4y = 8$

Solution 4.2

a) $x - y = -1$; $x + y = 3$ — Write each equation in $y = mx + b$ form.

$x - x - y = -1 - x$; $x - x + y = 3 - x$ — Subtract x from both sides

$-y = -x - 1$; $y = -x + 3$

$y = x + 1$ — Graph each equation using the slope-intercept form.

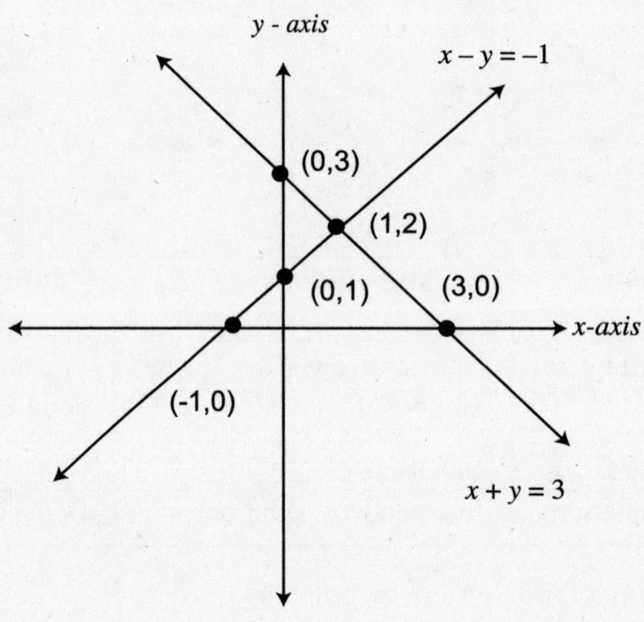

The graphs of the lines appear to intersect at (1, 2).

Check:

$x - y = -1$; $x + y = 3$ — Write the original equations.

Substitute $x = 1$, $y = 2$.

$1 - 2 = -1$; $1 + 2 = 3$ — True

Because (1,2) is a solution to both equations, (1, 2) is the solution to the system.

b) $y = -\frac{3}{2}x + 0$; $x + 2y = 4$	Write each equation in $y = mx + b$ form.
$x + 2y - x = 4 - x$	Subtract x from both sides.
$2y = -x + 4$	
$\frac{1}{2}(2y) = \frac{1}{2}(-x + 4)$	Multiply both sides by $\frac{1}{2}$.
$y = -\frac{1}{2}x + 2$	Graph each line using the slope-intercept form.

The graphs of the lines appear to intersect at (–2, 3).

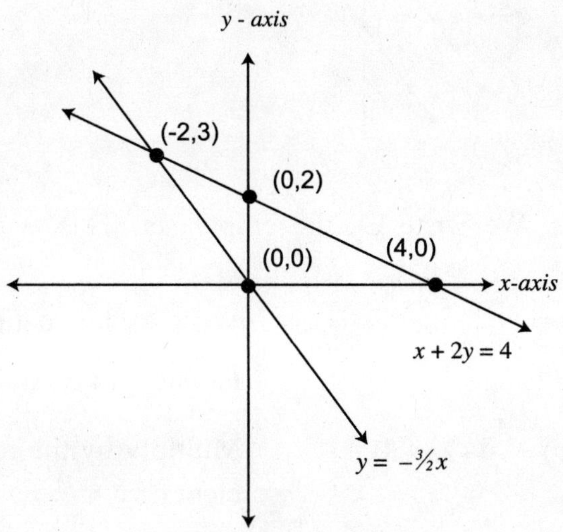

Check:

$y = -\frac{3}{2}x$; $x + 2y = 4$	Write the original equations.
$3 = -\frac{3}{2}(-2)$; $(-2) + 2(3) = 4$	Substitute $x = -2$, $y = 3$.
$3 = 3$; $4 = 4$	True

Because (–2, 3) is a solution to both equations, (–2, 3) is the solution to the system.

c) $3x - y = 2;\ 3x - y = -1$ — Write each equation in $y = mx + b$ form.

$3x - 3x - y = 2 - 3x;\ 3x - 3x - y = -1 - 3x$ — Subtract $3x$ from both sides.

$-y = -3x + 2;\ -y = -3x - 1$

$-1(-y) = -1(-3x + 2);\ -1(-y) = -1(-3x - 1)$ — Multiply by -1 to make the coefficient of y positive.

$y = 3x - 2;\ y = 3x + 1$ — Graph each line using the slope-intercept form.

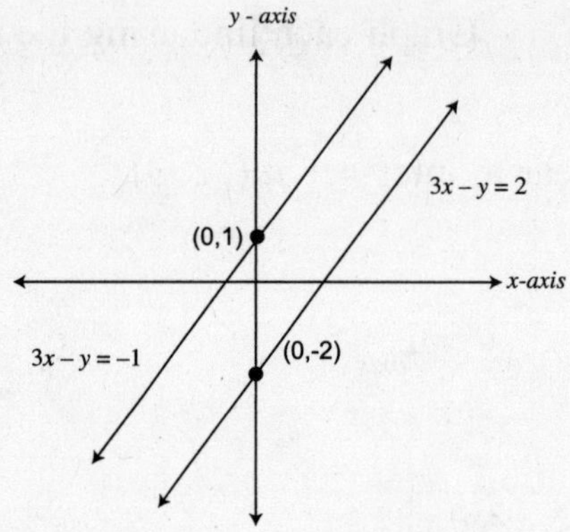

These are parallel lines. We write $\varnothing$, the empty set. This is called an **inconsistent system.**

d) $x + 2y = 4;\ 2x + 4y = 8$ — Write each equation in $y = mx + b$ form.

$2y = -x + 4;\ 4y = -2x + 8$ — Isolate y in each equation.

$\frac{1}{2}(2y) = \frac{1}{2}(-x + 4);\ \frac{1}{4}(4y) = \frac{1}{4}(-2x + 8)$ — Multiply by the reciprocal of the coefficient of y.

$y = -\frac{1}{2}x + 2\ ;y = -\frac{1}{2}x + 2$ — Graph each line using the slope-intercept form.

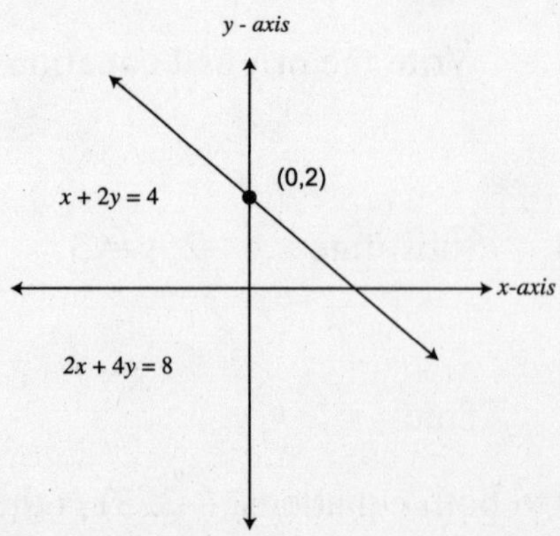

These are the same line. There are an infinite number of solutions. The equations are called **dependent.** The solution is every point on the line.

4.2 SOLVING SYSTEMS OF LINEAR EQUATIONS BY ADDITION

The solutions we found by graphing in the preceding section were integer solutions. It is certainly more difficult to find a solution like $\left(\frac{1}{9}, \frac{2}{3}\right)$ by graphing. Therefore, the next two sections present techniques for solving systems of linear equations without graphing.

SOLVING SYSTEMS OF LINEAR EQUATIONS BY ADDITION

To solve a system by addition:

1. Write each equation in the form $ax + by = c$. Multiply by the least common denominator (LCD) to eliminate any fractions.
2. If coefficients of either x or y are additive inverses (like 4 and –4), add the equations. If coefficients are not additive inverses, eliminate x's by multiplying each equation by the coefficient of x from the other equation. If the signs are the same, multiply one equation by –1.
3. Add the equations together.
4. Solve for y.
5. Substitute the value for y in either of the original equations to solve for x.
6. Check your answer.

Note: You may eliminate y's in the same manner, solve for x, substitute, and then solve for y. Depending on the coefficients of x and y, it is sometimes easier to eliminate y. Study the following exercises, keeping in mind that each can be solved by eliminating the other variable. This method is also called the **elimination method.**

Exercise 4.3

Solve each system:

a) $x + y = -5;\ 2x - y = 2$

b) $2x + y = 13;\ 3x - 2y = 2$

c) $2x + 5y = 27;\ 4x - 3y = 15$

d) $\frac{1}{2}x - \frac{1}{3}y = \frac{8}{3};\ \frac{1}{4}x + \frac{2}{3}y = -\frac{17}{6}$

Solution 4.3

a) $x + y = -5;\ 2x - y = 2$ — Because the coefficients of y are additive inverses (1 and –1), add the equations.

$$\begin{array}{r} x+y=-5 \\ 2x-y=\ \ 2 \\ \hline 3x \qquad =-3 \end{array}$$ Solve for x.

$x=-1$

$x+y=-5$ Use either original equation.

$(-1)+y=-5$ Substitute $x=-1$.

$-1+\mathbf{1}+y=-5+\mathbf{1}$ Add 1 to both sides.

$y=-4$

$(-1,-4)$ This is the proposed solution.

Check:

$x+y=-5$

$(-1)+(-4)=-5$ True

$2x-y=2$

$2(-1)-(-4)=2$ True

b) $2x+y=13$ Let's eliminate x.

$3x-2y=2$

$3(2x+y=13)$ Multiply by coefficients of x.

$\Rightarrow 6x+3y=39$

$2(3x-2y=2) \Rightarrow 6x-4y=4$

$-1(6x+3y=39)$ Because the signs are the same, multiply one equation by -1.

$\Rightarrow -6x-3y=-39$

$$\begin{array}{r} -6x-3y=-39 \\ 6x-4y=\ \ \ 4 \\ \hline -7y=-35 \end{array}$$ Add the equations.

$-\frac{1}{7}(-7y)=-\frac{1}{7}(-35)$ Solve for y.

$y=5$

$2x+y=13$ Use one of the original equations.

$2x+(5)=13$ Substitute $y=5$.

$2x+(5)-\mathbf{5}=13-\mathbf{5}$ Subtract 5 from both sides.

$2x=8$ Solve for x.

$x=4$

$(4,5)$ This is the proposed solution.

Check:

$2x + y = 13$

$2(4) + 5 = 13$ True

$3x - 2y = 2$

$3(4) - 2(5) = 2$ True

Note: We can combine two steps by multiplying by −3 and 2 or 3 and −2. If you had decided to eliminate y, the work would look like this:

$2(2x + y = 13) \Rightarrow 4x + 2y = 26$ Multiply by the coefficients of y.

$1(3x - 2y = 2) \Rightarrow 3x - 2y = 2$

$$\begin{array}{l} 4x + 2y = 26 \\ 3x - 2y = \ \ 2 \\ \hline 7x \qquad\ \ = 28 \end{array}$$

Add the equations.
Solve for x.

$\frac{7x}{7} = \frac{28}{7}$ Divide both sides by 7.

$x = 4$ Use one of the original equations.

$2x + y = 13$ Substitute $x = 4$.

$2(4) + y = 13$

$8 + y = 13$ Solve for y.

$8 - \mathbf{8} + y = 13 - \mathbf{8}$ Subtract 8 from both sides.

$y = 5$

(4, 5) is the same solution we obtained when we began by eliminating x. The check would be the same. Which way was easier?

c) $2x + 5y = 27$; $4x - 3y = 15$ Let's eliminate y. The signs are opposite, so there is no need to multiply by −1.

$3(2x + 5y = 27)$
$\Rightarrow 6x + 15y = 81$
$5(4x - 3y = 15)$
$\Rightarrow 20x - 15y = 75$

$$\begin{array}{l} \ \ 6x + 15y = \ 81 \\ 20x - 15y = \ 75 \\ \hline 26x \qquad\quad = 156 \end{array}$$

Add the equations.
Solve for x.

$x = 6$

$2x + 5y = 27$ Use one of the original equations.

$2(6) + 5y = 27$ Substitute $x = 6$.

$12 + 5y = 27$ Solve for y.

$12 - \mathbf{12} + 5y = 27 - \mathbf{12}$ — Subtract 12 from both sides.

$5y = 15$

$y = 3$

$(6, 3)$ — This is the proposed solution.

Check:

$2x + 5y = 27$

$2(6) + 5(3) = 27$ — True

$4x - 3y = 15$

$4(6) - 3(3) = 15$ — True

Therefore, (6, 3) is the solution.

d) $\frac{1}{2}x - \frac{1}{3}y = \frac{8}{3}$

$\frac{1}{4}x + \frac{2}{3}y = -\frac{17}{6}$

$6\left(\frac{1}{2}x - \frac{1}{3}y\right) = 6\left(\frac{8}{3}\right)$ — Multiply by the LCD to eliminate fractions.

$12\left(\frac{1}{4}x + \frac{2}{3}y\right) = 12\left(-\frac{17}{6}\right)$

$3x - 2y = 16$

$3x + 8y = -34$

$-1(3x - 2y = 16) = -3x + 2y = -16$ — Multiply by –1.

$$\begin{array}{r} -3x + 2y = -16 \\ 3x + 8y = -34 \\ \hline 10y = -50 \end{array}$$

Add the equations.
Solve for y.

$y = -5$

Rather than substitute $y = -5$ into one of the original equations, it may be easier to eliminate y and solve for x.

$4(3x - 2y = 16) \Rightarrow 12x - 8y = 64$

$3x + 8y = -34 \Rightarrow 3x + 8y = -34$ — Add the equations.

$15x = 30$ — Solve for x by dividing both sides by 15.

$x = 2$

$(2, -5)$ — This is the solution. It will check in both of the original equations.

DETERMINING WHEN A SYSTEM IS INCONSISTENT OR DEPENDENT

The graph of the solutions to the systems of parallel lines showed there were no solutions; similarly, the graph of a dependent system (coinciding lines) showed there were an infinite number of solutions. The following table will help you decide when a system is consistent and independent, inconsistent, or dependent when using the addition method:

Look For	Graph	Solution
Only one variable is eliminated.	Two intersecting lines	One ordered-pair solution
Both variables are eliminated and the resulting statement is true.	Lines coincide	Infinite number of solutions
Both variables are eliminated and the resulting statement is false.	Parallel lines	No solution or $\varnothing$

Exercise 4.4

Solve each system:

a) $2x - y = 6;\ 4x - 2y = 12$

b) $2x - 3y = 12;\ 2x - 3y = -3$

c) $x + 3y = 4;\ 2x + 6y = 8$

d) $2x + 4y = 0;\ 5x - 3y = 0$

Solution 4.4

a) $2x - y = 6;\ 4x - 2y = 12$

$-2(2x - y = 6)$ Eliminate y by multiplying the first equation by -2.
$\Rightarrow -4x + 2y = -12$

$$\begin{array}{r} -4x+2y=-12 \\ 4x-2y=\ \ 12 \\ \hline 0\ =\ \ \ 0 \end{array}$$

Add the equations.

Both variables were eliminated. The resulting statement is true. The lines coincide. There are an infinite number of solutions.

b) $2x - 3y = 12$

$2x - 3y = -3$ Coefficients of y will be additive inverses if we multiply one equation by -1.

$-1(2x - 3y = 12)$
$\Rightarrow -2x + 3y = -12$

$$\begin{array}{r} -2x+3y=-12 \\ 2x-3y=\ -3 \\ \hline 0=\ -15 \end{array}$$

Add the equations.
Both variables were eliminated. The resulting statement is false.

These are parallel lines. The solution set is empty, ∅.

c) $x + 3y = 4$

$2x + 6y = 8$

$-2(x + 3y = 4)$
$\Rightarrow -2x - 6y = -8$

Eliminate x by multiplying the first equation by –2.

$$\begin{array}{r} -2x-6y=-8 \\ 2x+6y=\ \ 8 \\ \hline 0=\ \ 0 \end{array}$$

Add the equations.
Both variables were eliminated, and the resulting statement is true.

The lines coincide. There are an infinite number of solutions.

d) $2x + 4y = 0$

$5x - 3y = 0$

Eliminate y by multiplying the first equation by 3 and the second equation by 4.

$3(2x + 4y = 0) \Rightarrow 6x + 12y = 0$
$4(5x - 3y = 0) \Rightarrow 20x - 12y = 0$

$$\begin{array}{r} 6x+12y=0 \\ 20x-12y=0 \\ \hline 26x=0 \end{array}$$

Add the equations.
Only one variable was eliminated.

$\frac{26x}{26} = \frac{0}{26}$

Solve for x by dividing both sides by 26.

$x = 0$

$2x + 4y = 0$ — Use one of the original equations.

$2(0) + 4y = 0$ — Substitute $x = 0$.

$0 + 4y = 0$ — Solve for y.

$\frac{4y}{4} = \frac{0}{4}$ — Divide both sides by 4.

$y = 0$

These lines intersect. There is one solution: (0, 0).

4.3 SOLVING SYSTEMS OF LINEAR EQUATIONS BY SUBSTITUTION

If one of the equations is not in $ax + by = c$ form, it is often more convenient to use the **substitution method** to solve a system.

1. Solve one equation for x or y, if necessary.
2. Substitute the expression from Step 1 into the other equation.
3. Solve the resulting equation.
4. Substitute the value from Step 3 into one of the original equations to find the other value.
5. Check, as necessary.

Exercise 4.5

Solve each system by substitution:

a) $2x + 3y = -12;\ y = 2x - 20$

b) $x + 3y = -14;\ -4x + 5y = -12$

c) $2x + 3y = 1;\ 4x + 5y = -1$

d) $x = \frac{1}{3}y + 1;\ 3x - y = -5$

e) $4x - y = 0;\ 8x - 2y = 0$

Solution 4.5

a) $2x + 3y = -12$

$y = 2x - 20$	This equation is solved for y.
$2x + 3(2x - 20) = -12$	Substitute $(2x - 20)$ for y.
$2x + 6x - 60 = -12$	Solve for x.
$8x - 60 + \mathbf{60} = -12 + \mathbf{60}$	Add 60 to both sides.
$8x = 48$	Divide both sides by 8.
$x = 6$	
Now solve for y:	
$y = 2x - 20$	It's easier to substitute into this equation because it is
$y = 2(6) - 20$	already solved for y, but you may use either equation.
$y = 12 - 20 = -8$	
$(6, -8)$	This is the solution.

Check:

$2x + 3y = -12$	
$y = 2x - 20$	These are the original equations.
$2(6) + 3(-8) = -12$	
$-8 = 2(6) - 20$	Substitute $x = 6$, $y = -8$.
$12 - 24 = -12$	
$-8 = 12 - 20$	Simplify each side.
$-12 = -12$	
$-8 = -8$	Both statements are true, so (6, –8) is the solution.
b) $x + 3y = -14$	The first equation can easily be solved for x
$-4x + 5y = -12$	because x has a coefficient of 1.
$x + 3y = -14$	Solve for x.
$x + 3y - 3y = -14 - 3y$	Subtract $3y$ from both sides.
$x = -3y - 14$	Now use substitution into the other equation.
$-4x + 5y = -12$	
$-4(-3y - 14) + 5y = -12$	Substitute $(-3y - 14)$ for x.
$12y + 56 + 5y = -12$	Solve for y.
$17y + 56 = -12$	
$17y + 56 - 56 = -12 - 56$	Subtract 56 from both sides.
$17y = -68$	Divide both sides by 17.
$y = -4$	
$x + 3y = -14$	Solve for x in the original equation.
$x + 3(-4) = -14$	Substitute $y = -4$.
$x - 12 = -14$	
$x - 12 + 12 = -14 + 12$	Add 12 to both sides.
$x = -2$	
(–2, –4)	This is the solution.

Check: Try the check on your own. Remember to check your answers in *both* of the original equations.

c) $2x + 3y = 1$
$4x + 5y = -1$

To use substitution, one equation must be solved for x or y. Let's solve the first equation for x.

$2x + 3y = 1$

$2x + 3y - 3y = 1 - 3y$ Subtract $3y$ from both sides.

$\frac{2x}{2} = \frac{-3y+1}{2}$ Divide both sides by 2. Use the commutative property to write $1 - 3y$ as $-3y + 1$.

$x = \frac{-3y+1}{2}$ Now substitute into the second equation.

$4x + 5y = -1$

$4\left(\frac{-3y+1}{2}\right) + 5y = -1$ Substitute $x = \frac{-3y+1}{2}$

$2(-3y + 1) + 5y = -1$ Solve for y.

$-6y + 2 + 5y = -1$ Combine similar terms.

$-y + 2 = -1$

$-y + 2 - 2 = -1 - 2$ Subtract 2 from both sides.

$-y = -3$

$-1(-y) = -1(-3)$ Multiply both sides by -1.

$y = 3$

Because $x = \frac{-3y+1}{2}$ Now solve for x.

$x = \frac{-3(3)+1}{2}$ Substitute $y = 3$.

$x = \frac{-9+1}{2}$

$x = -4$

$(-4, 3)$ This is the solution.

Check: Try the check on your own.

d) $x = \frac{1}{3}y + 1$
$3x - y = -5$ The first equation is already solved for x.

$3\left(\frac{1}{3}y + 1\right) - y = -5$ Substitute $\left(\frac{1}{3}y + 1\right)$ for x.

$y + 3 - y = -5$ Solve for y.

$3 = -5$ False

When both variables are eliminated and the resulting statement is false, the lines are parallel and the solution set is empty, $\varnothing$.

e) $4x - y = 0$

$8x - 2y = 0$ The first equation can easily be solved for y.

$4x - y = 0$

$4x - y - 4x = 0 - 4x$ Subtract $4x$ from both sides.

$-y = -4x$

$y = 4x$

$8x - 2y = 0$ Now use substitution.

$8x - 2(4x) = 0$ Substitute $4x$ for y.

$8x - 8x = 0$

$0 = 0$ True

When both variables are eliminated and the resulting statement is true, the lines coincide and there are an infinite number of solutions.

Although these systems can be solved with either the addition method or the substitution method, here are a few guidelines to help you decide which method to use:

* If one equation is already solved for one variable, or if one equation has a coefficient of 1, substitution works well.

* In all other cases, elimination is probably the better choice.

4.4 APPLICATIONS

The sample problems in this section are all application problems. Some can be solved using one variable; others require two variables (and, therefore, setting up two equations). The following steps will help you solve the two-variable problems in this section:

1. State what each variable you're using represents.
2. Translate the problem into a system of two equations.
3. Solve the system using addition or substitution.
4. Provide answers for the variable or variables as required.
5. Check your answer(s) in the original word problem.

NUMBER APPLICATIONS

You solved this type of problem in Chapter 3. You may want to review that chapter before starting these problems.

Exercise 4.6

One number is three times a second number. The sum of the numbers is 28. Find the numbers.

Solution 4.6

Let x = one number	State what each variable represents.
y = other number	
$x = 3y$	Write a system of equations.
$x + y = 28$	This system can be solved using substitution.
$(3y) + y = 28$	Substitute $x = 3y$ into the second equation.
$4y = 28$	Solve for y.
$\frac{4y}{4} = \frac{28}{4}$	Divide both sides by 4.
$y = 7$	Then solve $x = 3y$.
$x = 3(7)$	Substitute $y = 7$.
$x = 21$	

The two numbers are 7 and 21.

Check: 21 is three times 7.

21 + 7 is 28.

MONEY APPLICATIONS

An important concept used in both money and mixture applications involves finding the total value (in the money applications) and the total amount (in the mixture applications). If you purchase 10 stamps for 29¢ each, the total value of the stamps is found by multiplying the cost per stamp times the number of stamps: \$0.29(10) = \$2.90. Keep this idea in mind as you find the total value in the following exercises:

Exercise 4.7

525 tickets to the outdoor concert were sold. Reserved seats sold for \$10.00 and general admission tickets sold for \$7.50. The total receipts were \$4,250.00. How many of each type of ticket were sold?

Solution 4.7

Let x = the number of \$10.00 tickets sold	
y = the number of \$7.50 tickets sold	State what each variable represents.
$x + y = 525$	Total number of tickets sold.
$10.00x + 7.50y = 4{,}250.00$	Total value.
$x = 525 - y$	Solve the first equation for x.

$10.00(525 - y) + 7.50y$ $= 4{,}250.00$	Substitute into the second equation.
$5250.00 - 10.00y + 7.50y$ $= 4{,}250.00$	Solve for y.
$-2.50y = -1{,}000.00$	
$y = 400$	
$x + y = 525$	Solve for x.
$x + 400 = 525$	Substitute $y = 400$.
$x = 125$	
125 tickets sold for \$10.00 each	
400 tickets sold for \$ 7.50 each	
Check:	
$x + y = 525$	This is the original equation.
$(125) + (400) = 525$	Substitute $x = 125$, $y = 400$.
$525 = 525$	True
$10.00x + 7.50y = 4{,}250.00$	This is the original equation.
$10.00(125) + 7.50\,(400)$ $= 4{,}250.00$	Substitute $x = 125$, $y = 400$.
$1{,}250.00 + 3{,}000.00 = 4{,}250.00$	Multiply before adding.
$4{,}250.00 = 4{,}250.00$	True

MIXTURE APPLICATIONS

Mixture applications will often contain percentages that need to be changed to decimals. Note, for example, that an 80% alcohol solution contains 80% alcohol and 20% water. If there are 40 liters of this 80% alcohol solution, this means there are:

$40 \cdot 0.80$ liters of alcohol = 32 liters of alcohol and $40 \cdot 0.20$ liters of water = 8 liters of water

There may be other mixtures used in mixture problems, including antifreeze, potassium iodide, and disinfectant.

Exercise 4.8

A chemist needs 120 liters of 50% alcohol solution. The current stock contains two solutions: one is 80% alcohol, and the other is 30% alcohol. How many liters of each will be required to make the 120 liters of 50% alcohol solution?

Solution 4.8

Let x = the number of liters of 80% alcohol y = the number of liters of 30% alcohol	State what each variable represents.
$x + y = 120$	Total number of liters.
$0.80x + 0.30y = (0.50)120$	Total amount of alcohol.
$x = 120 - y$	Solve the first equation for x.
$0.80(120 - y) + 0.30y = 0.50(120)$	Substitute $120 - y$ for x.
$96 - 0.80y + 0.30y = 60$	Solve for y.
$96 - 0.50y = 60$	Isolate y.
$96 - 96 - 0.50y = 60 - 96$	Subtract 96 from both sides.
$-0.50y = -36$	Divide both sides by -0.50.
$y = 72$	
$x + y = 120$	Solve for x.
$x + 72 = 120$	Substitute $y = 72$.
$x = 48$	

48 liters of 80% alcohol must be mixed with 72 liters of 30% alcohol.

Check:

$48 + 72 = 120$	Substitute $x = 48$ and $y = 72$ into the original equations. Both are true statements.
$0.80(48) + 0.30(72) = 60$	

INTEREST APPLICATIONS

These types of applications generally involve money invested at different interest rates. Remember to convert an interest rate like 5% into a decimal (0.05) and multiply times the amount invested to find the amount of money made.

Exercise 4.9

Shawn inherited $12,000, which he wants to invest in two accounts, one that pays 4% interest and one that pays 6% interest. If he earns $660 in interest for the year, how much was invested at each rate?

Solution 4.9

Let x = the amount of money invested at 4% interest.

y = the amount of money invested at 6% interest.

$x + y = 12{,}000$	Write an equation to represent the total amount of money to invest.
$0.04x + 0.06y = 660$	Write an equation to represent the total interest earned. Convert each percent to a decimal. Multiply the decimal times the amount invested.

This system can be solved by addition or by substitution. Because we can easily solve for one variable in the first equation, we'll use substitution.

$y = 12{,}000 - x$	Solve for y.
$100[0.04x + 0.06(12{,}000 - x)]$ $= 100(660)$	Multiply both sides by 100 to clear the decimals.
$4x + 6(12{,}000 - x) = 66{,}000$	
$4x + 72{,}000 - 6x = 66{,}000$	Use the distributive property.
$-2x + 72{,}000 = 66{,}000$	Combine similar terms.
$-2x + 72{,}000 - 72{,}000$ $= 66{,}000 - 72{,}000$	Subtract 72,000 from both sides.
$-2x = -6{,}000$	Divide both sides by -2.
$x = 3{,}000$	

Be careful to answer the question. Because you are asked to find how much was invested at each rate, you must also solve for y:

$y = 12{,}000 - x = 12{,}000 - 3{,}000 = 9{,}000$

Thus, Shawn invested \$3,000 at 4% interest and \$9,000 at 6% interest.

4.5 SOLVING SYSTEMS OF LINEAR INEQUALITIES

You may want to review the "Graphing Linear Inequalities" section in Chapter 3.

To solve a system of linear inequalities by graphing:

1. Graph and shade each inequality on one set of axes. (Use a different type of shading or a different color for each inequality.)
2. Note that the solution to the system is where the shadings overlap.

Exercise 4.10

Graph the solution of each system:

a) $y \leq 2x + 1$; $y \geq -2x + 5$

b) $x + 2y < -4$; $y - 1 < x$

c) $x \geq 2$; $y < -3$

Solution 4.10

a) $y \le 2x + 1; y \ge -2x + 5$ — Both inequalities are in the form $y \,\square\, mx + b$.

$y \le 2x + 1$ — The line will be solid because the inequality symbol is $\le$. The shading will be below the line. Graph a solid line with slope $= 2 = \frac{2}{1}$ and y-intercept + 1.

$y \ge -2x + 5$ — The line will be solid because the inequality symbol is $\ge$. The shading will be above the line. Graph a solid line with slope $y \ge \frac{1}{2}x - 2$ and y-intercept +5.

The solution to the system is where the shadings overlap.

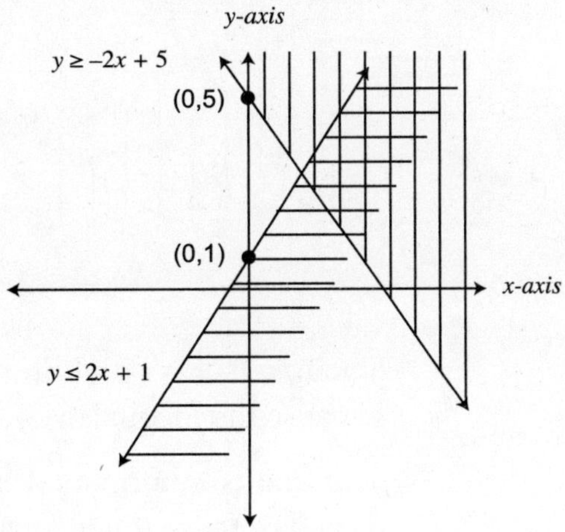

b) $x + 2y < -4; y - 1 < x$

First rewrite each inequality in $y \,\square\, mx + b$ form.

$x + 2y < -4; y - 1 < x$ — Get y alone on one side.

$x - x + 2y < -4 - x; y - 1 + 1 < x + 1$

$2y < -x - 4; y < x + 1$

$\frac{1}{2}(2y) < \frac{1}{2}(-x - 4)$

$y < -\frac{1}{2}x - 2$ — The line will be dotted because the inequality symbol is <. The shading will be below the line. Graph a dotted line with slope $= \frac{-1}{2}$ and y-intercept -2.

$y < x - 1$ — The line will be dotted because the inequality symbol is <. The shading will be below the line. Graph a dotted line with slope = $1 = \frac{1}{1}$ and y-intercept -1.

The solution to the system is where the shading overlaps.

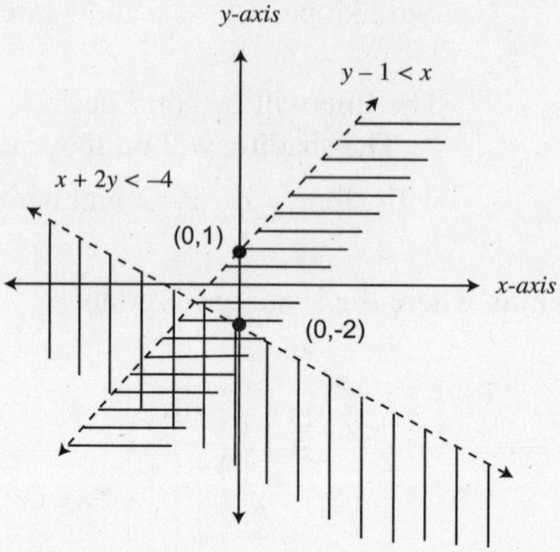

c) $x \geq 2$; $y < -3$

$x \geq 2$ — The line is a vertical line at $x = 2$. The line will be solid because the inequality is ≥. Shade to the right of the line.

$y < -3$ — The line is a horizontal line at $y = -3$. The line will be dotted because the inequality is <. Shade below the line.

The solution to the system is where the shadings overlap.

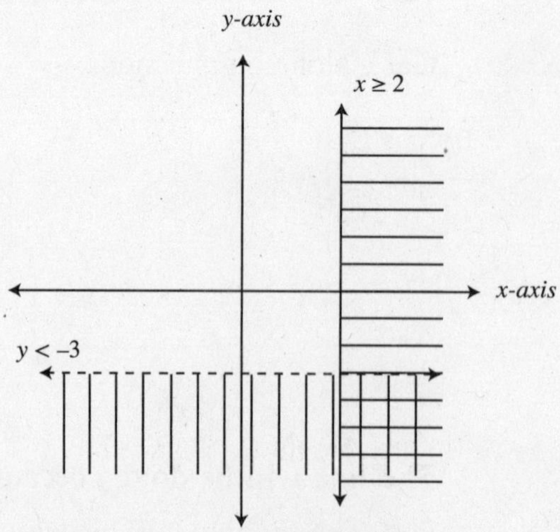

TEST YOURSELF

1) Determine whether the given ordered pair is a solution to the system:

(a) $(1, -2)$; $4x - 3y = 10$; $2x + y = 0$

(b) $(3, 4)$; $2x + y = 10$; $3x - 2y = 17$

(c) $(-3, 1)$; $4x - 5y = -17$; $3x - 2y = -11$

2) Solve each system by graphing:

(a) $3x + y = 13$; $x - 2y = -5$

(b) $3x - y = -3$; $x + 2y = -8$

(c) $4x - 3y = 8$; $8x - 6y = 16$

(d) $2x + y = 6$; $6x + 3y = 8$

3) Solve each system using the addition or elimination method:

(a) $2x + y = 3$; $x - y = 3$

(b) $3x - y = 2$; $6x + 2y = 20$

(c) $3x - 4y = 17$; $5x - 3y = 21$

(d) $10x - 6y = -2$; $15x + 9y = 9$

4) Solve each system:

(a) $x - 5y = 3$; $2x - 10y = 5$

(b) $2x + y = 6$; $4x + 2y = 12$

(c) $5x - y = 6$; $10x - 2y = 1$

5) Solve each system by substitution:

(a) $6x - 3y = -12$; $x = -2y - 7$

(b) $6x - 3y = -15$; $y = x + 4$

(c) $18x - y = 14$; $27x + 16y = -49$

(d) $3x + y = 4$; $6x + 2y = 8$

(e) $2x + y = 6$; $6x + 3y = 8$

6) Solve each problem:

(a) One number is one more than twice the second number. The sum of the numbers is 10. Find the numbers.

(b) The difference of two numbers is 8. The sum of the two numbers is 18. Find the numbers.

(c) One number is four times a second number. The sum of the numbers is 30. Find the numbers.

7) Solve each problem:

(a) There were 100 tickets sold to the puppet show. Children's tickets sold for $2.00, and adult tickets sold for $3.00. The total receipts were $265.00. How many of each type of ticket were sold?

(b) There were 126 tickets sold for the comedy show. Tickets for dinner and the show sold for $25.00, and tickets for the show only sold for $15.00. The total receipts were $2,790.00. How many of each type of ticket were sold?

(c) There were 425 tickets sold for the talent show. Students with an I.D. card could purchase tickets for $1.00. Other tickets sold for $3.50. The total receipts were $550.00. How many of each type of ticket were sold?

8) Solve each problem:

(a) A photographer needs 20 gallons of 25% acetic acid. She has 20% acetic acid and 40% acetic acid. How much of each will be needed?

(b) The garage needs 100 gallons of 40% antifreeze. The owner has 30% antifreeze and 70% antifreeze in stock. How many gallons of each type should he use?

(c) A piggy bank contained 113 coins in quarters and dimes for a total of $15.35. How many quarters and dimes were in the bank?

9) Solve each problem:

(a) Shawna won $6,000 and invested some at 3% interest and the rest at 5% interest. These investments earned $260 in a year. How much was invested at each rate?

(b) Brent has money invested in two accounts. He has part of the money invested at 8% and twice as much money invested at 7% interest. If the interest earned the first year is $880, how much did he invest in each account?

(c) Dr. Smith invested a sum of money at 5%. He invested $250 more than the first amount in an account that earned 6% interest. If his total annual interest income was $455, how much did he invest at each rate?

10) Graph the solutions of each system:

(a) $y \geq \frac{1}{2}x - 2$; $y < 3x + 1$

(b) $x + 3y \leq 6$; $y > x + 2$

(c) $x < 4$; $y \geq -2$

TEST YOURSELF ANSWERS

1) (a) Yes

(b) No

(c) Yes

2) (a) (3, 4)

(b) (−2, −3)

(c) Lines coincide; any ordered pair from the line is a solution.

(d) Parallel lines, ∅

3) (a) (2, −1)

(b) (2, 4)

(c) (3, −2)

(d) $\left(\frac{1}{5}, \frac{2}{3}\right)$

4) (a) Parallel lines, ∅

(b) Lines coincide; any ordered pair on the line is a solution

(c) Parallel lines, ∅

5) (a) (−3, −2)

(b) (−1, 3)

(c) $\left(\frac{5}{9}, -4\right)$

(d) Lines coincide; any ordered pair on the line is a solution.

(e) ∅

6) (a) 3 and 7

(b) 5 and 13

(c) 6 and 24

7) (a) 35 children, 65 adults

(b) 90 dinner and show, 36 show only

(c) 375 with I.D., 50 other

8) (a) 15 gal.@ 20%, 5 gal.@ 40%

(b) 75 gal. @ 30%,
25 gal. @ 70%

(c) 27 quarters, 86 dimes

9) (a) $4,000 at 5%, $2,000 at 3%

(b) $4,000 at 8%, $8,000 at 7%

(c) $4,000 at 5%, $4,250 at 6%

10) (a)

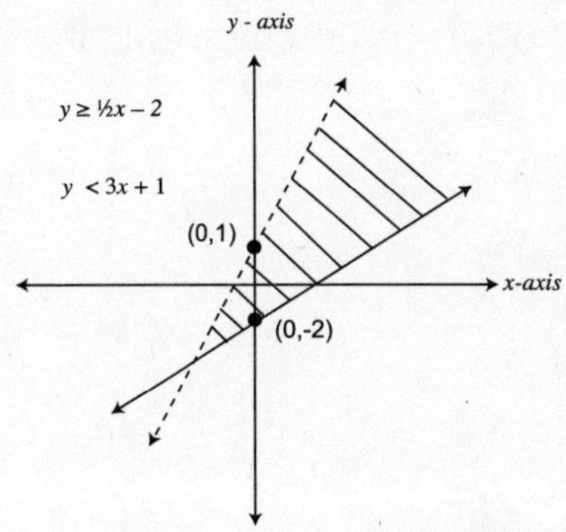

(b)

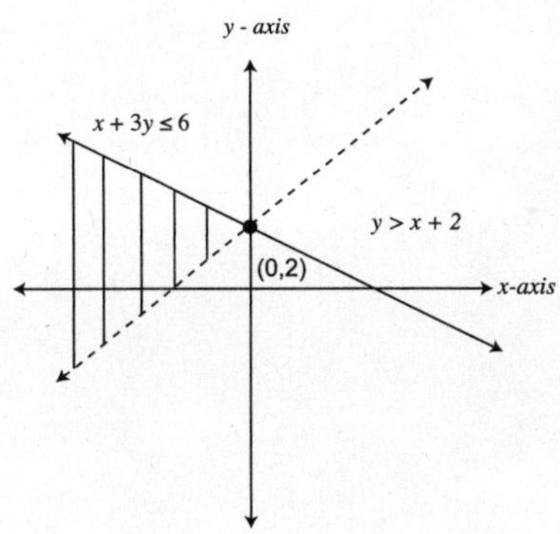

(c)

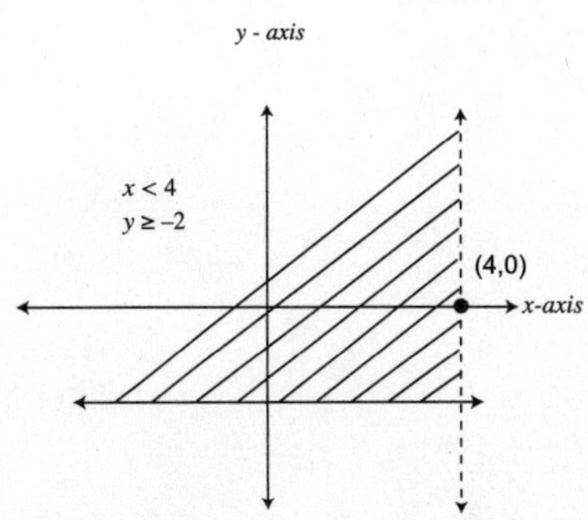

CHAPTER 5

Exponents and Polynomials

This chapter begins with a review of exponent properties for positive integers and then extends these properties to exponents that are rational numbers. This chapter also includes working with scientific notation (check your calculator for a button labeled EE). The chapter concludes with operations on polynomials including addition, subtraction, multiplication, and division by monomials and polynomials.

5.1 POSITIVE INTEGER EXPONENTS

Recall from Chapter 1 that $2^4 = 2 \cdot 2 \cdot 2 \cdot 2$, where 2 is called the **base**, and 4 is called the **exponent** or **power**.

Exercise 5.1

Simplify each expression. State the base and the exponent:

a) 3^2

b) -2^4

c) $(-2)^4$

Solution 5.1

a) $3^2 = 3 \cdot 3 = 9$ Base 3, exponent 2

b) $-2^4 = -1 \cdot 2 \cdot 2 \cdot 2 \cdot 2 = -16$ Base 2, exponent 4

c) $(-2)^4 = (-2)(-2)(-2)(-2) = 16$ Base –2, exponent 4

Be careful with exercises like (b) and (c). In (b), the base is 2, not (–2).

The following is a list of exponent laws you should memorize:

$a^m a^n = a^{m+n}$	When multiplying, if bases are the same (but only if they are the same), add the exponents. $a^m b^n$ is not $(ab)^{m+n}$.
$(a^m)^n = a^{mn}$	When raising a power to a power, multiply powers.
$(ab)^m = a^m b^m$	When raising a product to a power, raise each factor to the power.
$\left(\frac{a}{b}\right)^m = \frac{a^m}{b^m}, b \neq 0$	When raising a fraction to a power, raise the numerator and denominator to the power.

Exercise 5.2

Simplify each of the following expressions. Leave all answers in exponential form:

a) $4^2 \cdot 4^3$

b) $(-3)^3(-3)^4$

c) $x^4 \cdot x^2$

d) $y^3 \cdot y$

e) $a^5 \cdot a^3 \cdot a$

f) $3^{2a} \cdot 3^{3a}$

Solution 5.2

a) $4^2 \cdot 4^3 = 4^{2+3} = 4^5$	When multiplying and bases are equal, add the exponents.
b) $(-3)^3(-3)^4 = (-3)^{3+4} = (-3)^7$	Bases are equal, so add the exponents.
c) $x^4 \cdot x^2 = x^{4+2} = x^6$	Bases are equal, so add the exponents.
d) $y^3 \cdot y = y^3 \cdot y^1 = y^{3+1} = y^4$	Note that no written exponent means an exponent of 1: $y = y^1$.
e) $a^5 \cdot a^3 \cdot a = a^5 \cdot a^3 \cdot a^1$	$a = a^1$
$= a^{5+3+1}$	The bases are equal, so add the exponents.
$= a^9$	
f) $3^{2a} \cdot 3^{3a} = 3^{2a+3a}$	The bases are equal, so add the exponents.
$= 3^{5a}$	Combine similar terms: $2a + 3a = 5a$.

Exercise 5.3

Simplify each of the following expressions. Leave all answers in exponential form:

a) $(5^4)^2$

b) $(x^2)^3$

c) $(4^a)^3$

d) $(7^x)^y$

Solution 5.3

a) $(5^4)^2 = 5^{4\cdot 2} = 5^8$	To raise a power to a power, multiply the powers.
b) $(x^2)^3 = x^{2\cdot 3} = x^6$	To raise a power to a power, multiply the powers.
c) $(4^a)^3 = 4^{a\cdot 3}$	To raise a power to a power, multiply the powers.
$= 4^{3a}$	Use the commutative property: $a3 = 3a$.
d) $(7^x)^y = 7^{xy}$	To raise a power to a power, multiply the powers.

Exercise 5.4

Simplify each of the following expressions:

a) $(3y)^2$

b) $(-5^2)^4$

c) $(-2x^2y^3)^3$

d) $[(3a)^2]^3$

Solution 5.4

a) $(3y)^2 = 3^2\cdot y^2 = 9y^2$	To raise a product to a power, raise each factor to the power.
b) $(-5^2)^4 = (-1\cdot 5^2)^4$	First rewrite -5^2 as $-1\cdot 5^2$.
$= (-1)^4(5^2)^4$	To raise a product to a power, raise each factor to the power.
$=1\cdot 5^8 = 5^8$	$(-1)^4 = (-1)(-1)(-1)(-1) = 1$; $(5^2)^4 = 5^{2\cdot 4} = 5^8$.
c) $(-2x^2y^3)^3 = (-2)^3(x^2)^3(y^3)^3$	To raise a product to a power, raise each factor to the power.

$= -8x^6y^9$ — $(-2)^3 = -2 \cdot -2 \cdot -2 = -8$

d) $[(3a)^2]^3 = [(3^1a^1)^2]^3$ — Remember that $3 = 3^1$ and $a = a^1$.

$= [3^2a^2]^3$ — Work inside the parentheses first. To raise a product to a power, raise each factor to the power.

1.2×10^{18} — To raise a product to a power, raise each factor to the power. To raise a power to a power, multiply the powers.

$= 3^6a^6$

Exercise 5.5

Simplify each expression:

a) $\left(\frac{2}{5}\right)^3$

b) $\left(\frac{x}{y}\right)^4$

c) $\left(\frac{4c}{3b}\right)^3$

d) $\left(\frac{4m^2n}{p^3}\right)^2$

Solution 5.5

a) $\left(\frac{2}{5}\right)^3 = \frac{2^3}{5^3} = \frac{8}{125}$ — To raise a fraction to a power, raise the numerator and denominator to the power.

b) $\left(\frac{x}{y}\right)^4 = \frac{x^4}{y^4}$ — To raise a fraction to a power, raise the numerator and denominator to the power.

c) $\left(\frac{4c}{3b}\right)^3 = \frac{(4c)^3}{(3b)^3}$ — To raise a fraction to a power, raise the numerator and denominator to the power.

$= \frac{4^3c^3}{3^3b^3}$ — To raise a product to a power, raise each factor to the power.

$= \dfrac{64c^3}{27b^3}$ $\quad$ $4^3 = 4 \cdot 4 \cdot 4 = 64; 3^3 = 3 \cdot 3 \cdot 3 = 27.$

d) $\left(\dfrac{4m^2n}{p^3}\right)^2 = \dfrac{4^2\left(m^2\right)^2\left(n^1\right)^2}{\left(p^3\right)^2}$ $\quad$ To raise a fraction to a power, raise the numerator and denominator to the power. To raise a product to a power, raise each factor to the power.

$= \dfrac{16m^4n^2}{p^6}$ $\quad$ To raise a power to a power, multiply the powers.

Exercise 5.6

Simplify the following expressions:

a) $(5x^3)^2$

b) $(x^2)^3(x^4)^2$

c) $\left(\dfrac{2a^3}{b^2}\right)^4$

d) x^2y^2

Solution 5.6

a) $(5x^3)^2 = (5^1x^3)^2$ $\quad$ Write 5 as 5^1.

$= 5^{1\cdot2}x^{3\cdot2}$ $\quad$ To raise a product to a power, raise each factor to the power.

$= 25x^6$ $\quad$ $5^2 = 25$

b) $(x^2)^3(x^4)^2 = x^6 \cdot x^8$ $\quad$ To raise a power to a power, multiply the powers.

$= x^{6+8} = x^{14}$ $\quad$ The bases are equal, so add the exponents.

c) $\left(\dfrac{2a^3}{b^2}\right)^4 = \dfrac{(2a^3)^4}{\left(b^2\right)^4}$ $\quad$ To raise a fraction to a power, raise numerator and denominator to the power.

$= \dfrac{2^4a^{12}}{b^8}$ $\quad$ To raise a product to a power, multiply the powers.

$= \dfrac{16a^{12}}{b^8}$

d) x^2y^2 $\quad$ The bases are not the same. This is already in simplified form.

5.2 EXPONENT LAWS

The exponent laws you have previously applied to positive integers will now be expanded to include the use of zero as an exponent and negative integers as exponents.

ZERO AND NEGATIVE INTEGER EXPONENTS

We can extend our exponent laws to include all integers with the following definitions for zero and negative integer exponents:

Definition: $a^0 = 1$ for $a \neq 0$.

Definition: $a^{-n} = \frac{1}{a^n}$ for $a \neq 0$.

Exercise 5.7

Simplify the following expressions:

a) 9^0

b) $(-2x^3)^0$

c) $(4x)^0$

d) $4x^0$

Solution 5.7

a) $9^0 = 1$ Use the definition of 0 as an exponent.

b) $(-2x^3)^0 = 1$ Use the definition of 0 as an exponent.

c) $(4x)^0 = 1$ Use the definition of 0 as an exponent.

d) $4x^0 = 4 \cdot 1 = 4$ Use the definition of 0 as an exponent.

Note the difference between exercises (c) and (d). In (c), the product $4x$ is raised to the 0 power, while in (d) only x is raised to the 0 power.

Exercise 5.8

Evaluate each of the following expressions:

a) 4^{-2}

b) 2^{-3}

c) $(-2)^3$

d) $3x^{-4}$

e) $(2x)^{-3}$

Solution 5.8

a) $4^{-2} = \frac{1}{4^2} = \frac{1}{16}$ — Use the definition of a negative exponent.

b) $2^{-3} = \frac{1}{2^3} = \frac{1}{8}$ — Use the definition of a negative exponent.

c) $(-2)^3 = (-2)(-2)(-2)$ — The exponent is *not* negative. The base, –2, is negative.

$= -8$

d) $3x^{-4} = 3 \cdot \frac{1}{x^4}$ — Be careful here: Only the x is raised to a negative exponent, not the 3.

$= \frac{3}{x^4}$

e) $(2x)^{-3} = \frac{1}{(2x)^3}$ — Use the definition of a negative exponent. Because $(2x)$ is raised to the –3 power, $(2x)$ is moved to the denominator.

$= \frac{1}{2^3 x^3}$ — To raise a product to a power, raise each factor to the power.

$= \frac{1}{8x^3}$

EXPONENT LAW FOR DIVISION

We need to add one more law to our list of exponent laws. The exponent law for division is as follows: $\frac{a^m}{a^n} = a^{m-n}, a \neq 0$. That is, when dividing, if the bases are the same, subtract the exponent in the denominator from the exponent in the numerator.

Exercise 5.9

Simplify each of the following expressions. Write all answers with positive exponents:

a) $\frac{4^7}{4^5}$

b) $\frac{3}{3^5}$

c) $\dfrac{5^{-3}}{5^4}$

d) $\dfrac{x^7}{x^{-2}}$

Solution 5.9

a) $\dfrac{4^7}{4^5} = 4^{7-5} = 4^2 = 16$ Because the bases are equal, subtract the exponents.

b) $\dfrac{3}{3^5} = \dfrac{3^1}{3^5}$ Rewrite 3 as 3^1.

$= 3^{1-5}$ Because the bases are equal, subtract the exponents.

$= 3^{-4}$ You need to write the answer with a positive exponent. Use the definition of negative exponents to rewrite 3^{-4}.

$= \dfrac{1}{3^4} = \dfrac{1}{81}$

c) $\dfrac{5^{-3}}{5^4} = 5^{-3-4}$ Because the bases are equal, subtract the exponents.

$= 5^{-7}$ Use the definition of negative exponents to rewrite 5^{-7}.

$= \dfrac{1}{5^7}$

d) $\dfrac{x^7}{x^{-2}} = x^{7-(-2)}$ Because the bases are equal, subtract the exponents: $7 - (-2) = 7 + 2 = 9$

$= x^9$

Exercise 5.10

Simplify the following expressions. Write each answer with positive exponents:

a) $\dfrac{(6x^2)^3}{x^4}$

b) $\left(\dfrac{4a^{-3}}{a^2}\right)^2$

c) $\dfrac{\left(4^{-2}y^{-3}z^5\right)^{-2}}{\left(4y^{-2}z^2\right)^{-1}}$

d) $\dfrac{(3a)^{-3}}{(3a)^{-5}}$

e) $\dfrac{(y^3)^5(y^2)^{-4}}{(y^{-3})^3}$

Solution 5.10

a) $\dfrac{\left(6x^2\right)^3}{x^4} = \dfrac{6^3x^6}{x^4}$ — To raise a product to a power, raise each factor to the power.

$= 216x^{6-4}$ — Because the bases are equal, subtract the exponents.

$= 216x^2$

b) $\left(\dfrac{4a^{-3}}{a^2}\right)^2 = \left(4a^{-3-2}\right)^2$ — Simplify inside the parentheses first.

$= (4a^{-5})^2$

$= 4^2a^{-5\cdot 2}$ — To raise a product to a power, raise each factor to the power.

$=16a^{-10} = \dfrac{16}{a^{10}}$ — Use $a^{-10} = \dfrac{1}{a^{10}}$.

c) $\dfrac{\left(4^{-2}y^{-3}z^5\right)^{-2}}{\left(4y^{-2}z^2\right)^{-1}} = \dfrac{4^{-2(-2)}y^{-3(-2)}z^{5(-2)}}{4^{1(-1)}y^{-2(-1)}z^{2(-1)}}$ — A fraction bar is a grouping symbol. You must simplify above the bar and below the bar separately.

To raise a product to a power, raise each factor to the power.

$= \dfrac{4^4 y^6 z^{-10}}{4^{-1} y^2 z^{-2}}$

$= 4^{4-(-1)} y^{6-2} z^{-10-(-2)}$ To divide when the bases are equal, subtract the exponents.

$= 4^5 y^4 z^{-8}$ To write the answer with positive exponents, use the definition of negative exponents.

$= \dfrac{4^5 y^4}{z^8}$ Use $z^{-8} = \dfrac{1}{z^8}$.

d) $\dfrac{(3a)^{-3}}{(3a)^{-5}} = (3a)^{-3-(-5)}$ The bases are equal, so subtract the exponents.

$= (3a)^2$ $-3 - (-5) = -3 + 5 = 2$

$= 9a^2$ To raise a power to a power, raise each factor to the power.

e) $\dfrac{(y^3)^5 (y^2)^{-4}}{(y^{-3})^3} = \dfrac{y^{15} y^{-8}}{y^{-9}}$ To raise a product to a power, raise each factor to the power.

$= \dfrac{y^{15-8}}{y^{-9}}$ Simplify the numerator.

$= \dfrac{y^7}{y^{-9}} = y^{7-(-9)} = y^{16}$ The bases are equal, so subtract the exponents and simplify.

The following section summarizes the exponent laws and definitions we have covered in this chapter.

SUMMARY OF EXPONENT LAWS

$a^m \cdot a^n = a^{m+n}$ When multiplying, if the bases are equal, add the exponents.

$(a^m)^n = a^{mn}$ When raising a power to a power, multiply the powers.

$(ab)^m = a^m b^m$ When raising a product to a power, raise each factor to the power.

$\left(\dfrac{a}{b}\right)^m = \dfrac{a^m}{b^m}, b \neq 0$ When raising a fraction to a power, raise the numerator and the denominator to the power.

$\dfrac{a^m}{a^n} = a^{m-n}, a \neq 0$ When dividing, if the bases are equal, subtract the exponent in the denominator from the exponent in the numerator.

$a^0 = 1$ for $a \neq 0$	A nonzero number raised to the zero power equals 1.
$a^{-n} = \dfrac{1}{a^n}$ for $a \neq 0$.	A nonzero number raised to the $-n$ power equals 1 divided by the number raised to the n power.

Exercise 5.11

Simplify. Any answers that contain exponents should be written with positive exponents only:

a) $x^5 \cdot x^2 \cdot x$

b) $(4a^2b^3)^2(2ab^4)^3$

c) $(-5x^2y^3)^{-2}$

d) $\dfrac{8x^0y}{2y^0}$

e) $\dfrac{(8ab^2)^2(4a^3b^3)}{(4a^3b^3)(2a^6b)}$

f) $\dfrac{x^{-3}}{(x^{-2})^2(x^{-5})^{-3}}$

Solution 5.11

a) $x^5 \cdot x^2 \cdot x = x^{5+2+1} = x^8$ — The bases are equal, so add the exponents. Remember that $x = x^1$.

b) $(4a^2b^3)^2(2ab^4)^3$

$= (4^2a^4b^6)(2^3a^3b^{12})$ — To raise a product to a power, multiply the powers.

$= (16a^4b^6)(8a^3b^{12})$ — Multiply.

$= (16 \cdot 8a^{4+3}b^{6+12})$ — Multiply the coefficients. Add the exponents on equal bases.

$= 128a^7b^{18}$

c) $(-5x^2y^3)^{-2}$

$= \dfrac{1}{(-5x^2y^3)^2}$ — Use the definition of a negative exponent.

$= \dfrac{1}{(-5)^2x^{2\cdot2}y^{3\cdot2}}$ — Raise each factor to the power.

$$=\frac{1}{25x^4y^6}$$

d) $\frac{8x^0y}{2y^0}=\frac{8y}{2}=4y$

Use the definition of zero as an exponent, so $x^0 = 1$, $y^0 = 1$. Divide the coefficients.

e) $\frac{(8ab^2)^2(4a^3b^3)}{(4a^3b^3)(2a^6b)}$

$$=\frac{(64a^2b^4)(4a^3b^3)}{(4a^3b^3)(2a^6b)}$$

To raise a product to a power, multiply the powers.

$$=\frac{256a^5b^7}{8a^9b^4}$$

To multiply, add the exponents on equal bases.

$$=32a^{5-9}b^{7-4}$$

Divide the coefficients, subtract the exponents on equal bases.

$$=32a^{-4}b^3=\frac{32b^3}{a^4}$$

Use the definition of negative exponent, $a^{-4}=\frac{1}{a^4}$.

f) $\frac{x^{-3}}{(x^{-2})^2(x^{-5})^{-3}}$

$$=\frac{x^{-3}}{x^{-4}x^{15}}$$

To raise a power to a power, multiply the powers.

$$=\frac{x^{-3}}{x^{-4+15}}$$

To multiply with equal bases, add the exponents.

$$=\frac{x^{-3}}{x^{11}}$$

$$=x^{-3-11}=x^{-14}=\frac{1}{x^{14}}$$

To divide, subtract the exponents, and then use the definition of a negative exponent.

5.3 SCIENTIFIC NOTATION

Scientific notation is a shorthand method used for writing very large and very small numbers.

WRITING NUMBERS IN SCIENTIFIC NOTATION

To write numbers in scientific notation, do the following:

1. Move the decimal point to the right of the first nonzero digit. Count the number of places you moved the decimal point.

2. Multiply the number from Step 1 times 10 raised to a power:

   + the number of places you moved the decimal point if you moved the decimal to the left.

   – the number of places you moved the decimal point if you moved the decimal to the right.

Exercise 5.12

Write each number in scientific notation:

a) 247,000

b) 16

c) 3,000,421,000

Solution 5.12

a) 247,000	This is the given number.
2.47000	Move the decimal to the right of 2.
	The decimal point moved 5 places to the left.
$= 2.47 \times 10^{+5}$	Use +5 as the exponent of 10, because the decimal was moved to the left 5 places.
b) 16	This is the given number.
1.6	Move the decimal to the right of 1.
	The decimal point moved 1 place to the left.
$= 1.6 \times 10^{+1}$	Use +1 as the exponent of 10, because the decimal point was moved to the left 1 place.
or	
$= 1.6 \times 10$	$10^{+1} = 10^1 = 10$; you can use any of these forms.
c) 3,000,421,000	This is the given number.
3.000421000	Move the decimal to the right of 3.
	The decimal point moved 9 places to the left.
$= 3.000421 \times 10^9$	Use +9 or 9 as the exponent of 10, because the decimal point was moved to the left 9 places.

Exercise 5.13

Write each number in scientific notation:

a) 0.00562

b) 0.0000000048

c) 0.030056

Solution 5.13

a) 0.00562	This is the given number.
005.62	Move the decimal to the right of 5.
	The decimal point moved 3 places to the right.
$= 5.62 \times 10^{-3}$	Use –3 as the exponent of 10, because the decimal point was moved to the right 3 places.
b) 0.0000000048	This is the given number.
000000004.8	Move the decimal to the right of 4.
	The decimal point moved 9 places to the right.
$= 4.8 \times 10^{-9}$	Use –9 as the exponent of 10, because the decimal point was moved to the right 9 places.
c) 0.030056	This is the given number.
03.0056	Move the decimal to the right of 3.
	The decimal point moved 2 places to the right.
$= 3.0056 \times 10^{-2}$	Use –2 as the exponent of 10, because the decimal point was moved to the right 2 places.

WRITING NUMBERS IN EXPANDED FORM

To write a number in expanded form:

1. If the exponent of 10 is positive (+), move the decimal point to the right the same number of places as the exponent. Add zeros as necessary.
2. If the exponent of 10 is negative (–), move the decimal point to the left the same number of places as the exponent. Add zeros as necessary.

Exercise 5.14

Write each number in expanded form:

a) 6.28×10^3

b) 2.15×10^6

c) 4.5896×10^2

Solution 5.14

a) 6.28×10^3	This is the given number.
628 __ .	Move the decimal point 3 places to the right.
6,280	Add necessary 0.
= 6,280	This is the answer.

b) 2.15×10^6	This is the given number.
215 __ __ __ __ .	Move the decimal point 6 places to the right.
2,150,000	Add the necessary zeros.
= 2,150,000	This is the answer.
c) 4.5896×10^2	This is the given number.
458.96	Move the decimal point 2 places to the right.
= 458.96	This is the answer.

Exercise 5.15

Write each number in expanded form:

a) 3.79×10^{-5}
b) 2.6×10^{-3}
c) 1.57896×10^{-1}

Solution 5.15

a) 3.79×10^{-5}	This is the given number.
0. __ __ __ __ 379	Move the decimal point 5 places to the left.
0.0000379	Add necessary zeros.
= 0.0000379	This is the answer. We usually write a 0 to the left of the decimal point.
b) 2.6×10^{-3}	This is the given number.
0. __ __ 26	Move the decimal point 3 places to the left.
0.0026	Add necessary zeros.
= 0.0026	We usually write a 0 to the left of the decimal point.
c) 1.57896×10^{-1}	This is the given number.
.157896	Move the decimal point 1 place to the left.
= 0.157896	We usually write a 0 to the left of the decimal point.

USING EXPONENT LAWS WITH SCIENTIFIC NOTATION

When numbers are written in scientific notation, you can use the exponent laws to simplify products and quotients. Remember that $a^m a^n = a^{m+n}$ and $\left(\frac{a^m}{a^n}\right) = a^{m-n}$ can be used only when bases are the same.

Exercise 5.16

Find the product. Write answers in scientific notation:

a) $(2 \times 10^6) \times (8 \times 10^4)$
b) $(5 \times 10^3) \times (3 \times 10^{-5})$

c) $(4 \times 10^{-2}) \times (9 \times 10^{-4})$

d) $(-3 \times 10^{5}) \times (-8 \times 10^{-6})$

Solution 5.16

a) $(2 \times 10^{6}) \times (8 \times 10^{4})$

$= 2 \times 8 \times 10^{6} \times 10^{4}$	Use the commutative and associative properties.
$= 16 \times 10^{6+4}$	The bases are equal, so add the exponents.
$= 16 \times 10^{10}$	You're not finished yet! Now convert to scientific notation.
$= (1.6 \times 10^{1}) \times 10^{10}$	$16 = 16. = 1.6 \times 10^{1}$
$= 1.6 \times 10^{11}$	The bases are equal, so add the exponents.

b) $(5 \times 10^{3}) \times (3 \times 10^{-5})$

$= 5 \times 3 \times 10^{3} \times 10^{-5}$	
$= 15 \times 10^{3+(-5)}$	The bases are equal, so add the exponents.
$= 15 \times 10^{-2}$	You're not finished yet! Now convert to scientific notation.
$= 15. \times 10^{-2}$	$15 = 15. = 1.5 \times 10^{1}$
$= (1.5 \times 10^{1}) \times 10^{-2}$	$10^{1} \times 10^{-2} = 10^{1+(-2)}$
$= 1.5 \times 10^{-1}$	

c) $(4 \times 10^{-2}) \times (9 \times 10^{-4})$

$= 4 \times 9 \times 10^{-2} \times 10^{-4}$	
$= 36 \times 10^{-2+(-4)}$	The bases are equal, so add the exponents.
$= 36 \times 10^{-6}$	You're not finished yet! Now convert to scientific notation.
$= (3.6 \times 10^{1}) \times 10^{-6}$	$36 = 36. = 3.6 \times 10^{1}$
$= 3.6 \times 10^{1+(-6)}$	The bases are equal, so add the exponents.
$= 3.6 \times 10^{-5}$	

d) $(-3 \times 10^{5}) \times (-8 \times 10^{-6})$

$= (-3) \times (-8) \times 10^{5} \times 10^{-6}$	
$= 24 \times 10^{5+(-6)}$	The bases are equal, so add the exponents.
$= 24 \times 10^{-1}$	Now convert to scientific notation.
$= (2.4 \times 10^{1}) \times 10^{-1}$	$24 = 24. = 2.4 \times 10$
$= 2.4 \times 10^{1+(-1)}$	The bases are equal, so add the exponents.
$= 2.4 \times 10^{0}$	

Exercise 5.17

Find the quotient. Write answers in scientific notation:

a) $\dfrac{12\times10^{6}}{4\times10^{-3}}$

b) $\dfrac{3.6\times10^{-2}}{6\times10^{3}}$

c) $\dfrac{-4.8\times10^{-7}}{-2.4\times10^{-4}}$

Solution 5.17

a) $\dfrac{12\times10^{6}}{4\times10^{-3}}$

$=\dfrac{12}{4}\times10^{6-(-3)}$ The bases are equal, so subtract the exponents.

$=3\times10^{9}$ $6-(-3)=6+3=9$

b) $\dfrac{3.6\times10^{-2}}{6\times10^{3}}$

$=\dfrac{3.6}{6}\times10^{-2-3}$ The bases are equal, so subtract the exponents.

$= 0.6 \times 10^{-5}$ You're not finished yet! Now convert to scientific notation.

$= (6 \times 10^{-1}) \times 10^{-5}$ $0.6 = 6. \times 10^{-1}$.

$= 6 \times 10^{-6}$ $10^{-1} \times 10^{-5} = 10^{-1+(-5)} = 10^{-6}$

c) $\dfrac{-4.8\times10^{-7}}{-2.4\times10^{-4}}$

$=\dfrac{-4.8}{-2.4}\times10^{-7-(-4)}$ The bases are equal, so subtract the exponents.

$= 2 \times 10^{-3}$ $-7-(-4)=-7+4=-3$

Exercise 5.18

Simplify each expression. Write all answers in scientific notation:

a) $\dfrac{(4\times10^3)(5\times10^6)}{2\times10^3}$

b) $\dfrac{(3\times10^{-4})(6\times10^{-5})}{9\times10^3}$

c) $\dfrac{(5\times10^3)(8\times10^{-2})}{2\times10^4}$

Solution 5.18

a) $\dfrac{(4\times10^3)(5\times10^6)}{2\times10^3}$

$=\dfrac{4\times5\times10^{6+3}}{2\times10^3}$ To multiply with equal bases, add the exponents.

$=\dfrac{20\times10^9}{2\times10^3}$ Simplify.

$=\dfrac{20}{2}\times10^{9-3}$ Divide the coefficients. Subtract the exponents for equal bases.

$= 10 \times 10^6$ Don't stop yet. Write 10 in scientific notation.

$=1 \times 10^1 \times 10^6$ Write 10 as 1×10^1.

$=1 \times 10^{1+6} = 1 \times 10^7$ To multiply with equal bases, add the exponents.

b) $\dfrac{(3\times10^{-4})(6\times10^{-5})}{9\times10^3}$

$=\dfrac{3\times6\times10^{-4}\times10^{-5}}{9\times10^3}$

$=\dfrac{18\times10^{-4+(-5)}}{9\times10^3}$ Multiply. Add the exponents when working with equal bases.

$=\dfrac{18}{9}\times10^{-9-3}$ Divide the coefficients. Subtract the exponents when working with equal bases.

$= 2 \times 10^{-12}$

c) $\dfrac{(5\times10^3)(8\times10^{-2})}{2\times10^4}$

$=\dfrac{5\times8\times10^{3+(-2)}}{2\times10^4}$ Multiply. Add the exponents on equal bases.

$=\dfrac{40\times10^1}{2\times10^4}$ Simplify.

$=\dfrac{40}{2}\times10^{1-4}$ Divide the coefficients. Subtract the exponents on equal bases.

$= 20 \times 10^{-3}$ Don't stop yet. Write 20 in scientific notation.

$= 2 \times 10^1 \times 10^{-3}$

$= 2 \times 10^{1+(-3)} = 2 \times 10^{-2}$ To multiply with equal bases, add the exponents.

5.4 ADDING AND SUBTRACTING POLYNOMIALS

This section presents the language and some of the techniques used to add and subtract polynomials. Polynomials play such a large part in the study of algebra that understanding the language used with polynomials is essential to the remainder of text.

VOCABULARY FOR POLYNOMIALS

The following table presents some language you need to understand the remaining sections of this chapter:

Vocabulary	Meaning	Example
Term	A number or a product of a number and one or more variables	In $6x^2 + 5x + 8$ there are 3 **terms:** $6x^2$, $5x$, and 8.
Numerical coefficient	Numerical factor in a term	4 is the **numerical coefficient** in $4x^3$. -2 is the **numerical coefficient** in $-2xy^2$. 1 is the **numerical coefficient** in xy.
Polynomial	Expression containing a finite sum of terms	$4x$ $x^2 + 2x - 1$
Monomial	Polynomial with 1 term	6 $2x$
Binomial	Polynomial with 2 terms	$x^2 - 9$ $2x + 6$
Trinomial	Polynomial with 3 terms	$x^2 + 2x + 1$
Degree of a term	Sum of the exponents of the variables within the term	**Degree** of $6x^2y^3$ is 5. **Degree** of $2x = 2x^1$ is 1. **Degree** of $5 = 5x^0$ is 0.
Degree of a polynomial in 1 variable	Highest degree of any term	Degree of $4x^3 - 2x$ is 3. Degree of $x^5 - 2x^3 + 6x$ is 5.
Similar terms or like terms	Terms that have the same variables and exponents	$4x^3$ and $-6x^3$ $6a$ and $2a$

Exercise 5.19

Identify whether each of the following polynomials is a monomial, binomial, or trinomial:

a) $4x + 6$

b) $7x^2 - 5xy + 3y^4$

c) $5xyz$

Solution 5.19

a) $4x + 6$ is a binomial.

b) $7x^2 - 5xy + 3y^4$ is a trinomial.

c) $5xyz$ is a monomial.

Exercise 5.20

Find the degree of each polynomial:

a) $4x$

b) 6

c) $5xyz$

d) $7x^2 - 5xy + 3y^4$

Solution 5.20

a) $4x = 4x^1$ has degree 1 — The degree of a term is the sum of the exponents of the variables.

b) $6 = 6x^0$ has degree 0 — The degree of any constant will be 0, because the constant can be written as the constant times x^0.

c) $5xyz = 5x^1y^1z^1$ has degree 3 — The degree of a term is the sum of the exponents of the variables.

d) To find the degree of a polynomial, first find the degree of each term:

$7x^2$ has degree 2

$5xy$ has degree 2 — $5xy = 5x^1y^1$.

$3y^4$ has degree 4

The degree of the polynomial is the highest degree of any term, so the degree of $7x^2 - 5xy + 3y^4$ is 4.

ADDING POLYNOMIALS

Recall from Chapter 2 that only similar terms can be added or subtracted. The following steps help you add polynomials:

1. Use the definition of subtraction to change subtraction into addition of the opposite.
2. Group similar terms together, making use of the commutative and associative properties.
3. Add the numerical coefficients of similar terms.

Exercise 5.21

Add:

a) $(3x^2 + 4x + 7) + (5x^2 + 2x + 12)$

b) $(4x^2 - 6x - 3) + (6x^2 + 8x - 5)$

c) $-2y^3 + 8y - 4 + 7y^3 - y^2 + 5$

Solution 5.21

a) $(3x^2 + 4x + 7) + (5x^2 + 2x + 12)$

$= (3x^2 + 5x^2) + (4x + 2x) + (7 + 12)$ — Group similar terms together using the comutative and associative properties.

$= (3 + 5)x^2 + (4 + 2)x + (19)$ — Add the numerical coefficients of similar terms.

$= 8x^2 + 6x + 19$

b) $(4x^2 - 6x - 3) + (6x^2 + 8x - 5)$

$= (4x^2 + (-6x) + (-3)) + 6x^2 + 8x + (-5)$ — Rewrite each polynomial using the definition of subtraction to change subtraction to addition of the opposite.

$= (4x^2 + 6x^2) + (-6x + 8x) + (-3 + (-5))$ — Group similar terms together.

$= 10x^2 + 2x + (-8)$ — Add numerical coefficients of similar terms.

$= 10x^2 + 2x - 8$ — $+ (-8)$ is usually written -8.

c) $-2y^3 + 8y - 4 + 7y^3 - y^2 + 5$

$= -2y^3 + 8y + (-4) + 7y^3 + (-1y^2) + 5$ — Use the definition of subtraction.

$= (-2y^3 + 7y^3) - y^2 + 8y + (-4 + 5)$ — Group similar terms together.

$= 5y^3 - y^2 + 8y + 1$ — Add numerical coefficients of similar terms.

SUBTRACTING POLYNOMIALS

To subtract two polynomials, change all the signs of the polynomial behind the subtraction sign, and then add the polynomials.

Exercise 5.22

Subtract:

a) $(4y^2 - y + 6) - (2y^2 + 5y - 8)$

b) $(5a^4 - 3a^3 - 8) - (-2a^4 + 7a^3 - 2)$

Solution 5.22

a) $(4y^2 - y + 6) - (2y^2 + 5y - 8)$

$= (4y^2 - y + 6) + (-2y^2 - 5y + 8)$	Change all signs in the second polynomial.
$= (4y^2 - 2y^2) + (-y - 5y) + (6 + 8)$	Group similar terms.
$= 2y^2 - 6y + 14$	Combine similar terms.

b) $(5a^4 - 3a^3 - 8) - (-2a^4 + 7a^3 - 2)$

$= (5a^4 - 3a^3 - 8) + (2a^4 - 7a^3 + 2)$	Change all signs in the second polynomial.
$= (5a^4 + 2a^4) + (-3a^3 - 7a^3) + (-8 + 2)$	Group similar terms.
$= 7a^4 - 10a^3 - 6$	Combine similar terms.

Exercise 5.23

Subtract $6x^2 + 5x - 8$ from $7x^2 - 8$.

Solution 5.23

$(7x^2 - 8) - (6x^2 + 5x - 8)$	Note that the subtraction sign goes in front of $6x^2 + 5x - 8$.
$= (7x^2 - 8) + (-6x^2 - 5x + 8)$	Change all signs in the second polynomial.
$= (7x^2 - 6x^2) - 5x + (-8 + 8)$	Group similar terms.
$= x^2 - 5x$	Combine similar terms.

5.5 MULTIPLYING POLYNOMIALS

We begin the topic of multiplication of polynomials by multiplying monomials (one-term polynomials), then polynomials.

MULTIPLYING MONOMIALS BY MONOMIALS

To multiply a monomial times a monomial, multiply the coefficients and add the exponents on equal bases.

Exercise 5.24

Multiply:

a) $(4x^2)(-2x^3)$

b) $(3x)(5x)$

c) $y^2(6y^3)$

d) $(-2m^2)(3n^2)$

e) $(4x^2y)(2y^3)(5x^4)$

Solution 5.24

a) $(4x^2)(-2x^3) = (4)(-2)x^{2+3}$ — Multiply the coefficients. Add the exponents on equal bases.

$= -8x^5$

b) $(3x)(5x) = (3x^1)(5x^1)$ — Rewrite each variable with an exponent: $x = x^1$.

$= (3)(5)x^{1+1}$ — Multiply the coefficients. Add the exponents on equal bases.

$= 15x^2$

c) $y^2(6y^3) = (1)(6)y^{2+3}$ — $y^2 = 1y^2$

$= 6y^5$

d) $(-2m^2)(3n^2) = (-2)(3)m^2n^2$ — Multiply the coefficients. Bases are not equal, so exponents cannot be added.

$= -6m^2n^2$

e) $(4x^2y)(2y^3)(5x^4) = (4)(2)(5)x^{2+4}y^{1+3}$ — Multiply the coefficients. Add the exponents on equal bases.

$= 40x^6y^4$

MULTIPLYING TWO POLYNOMIALS

To multiply two polynomials:

1. Multiply each term of the second polynomial by each term of the first polynomial.
2. Add any similar terms.

Exercise 5.25

Multiply:

a) $2y(3y^4 + 4)$

b) $4p(3p^2 - 6p + 2)$

c) $(2a + 4)(3a + 5)$

d) $(3x + 5)(2x - 3)$

e) $(4b - 3)(2b - 7)$

Solution 5.25

a) $2y(3y^4 + 4) = 2y(3y^4) + 2y(4)$ — Multiply each term of the second polynomial by $2y$.

$= 6y^5 + 8y$	Remember to use $2y = 2y^1$. Multiply the coefficients. Add the exponents on equal bases.
b) $4p(3p^2 - 6p + 2)$	
$= 4p(3p^2) + 4p(-6p) + 4p(2)$	Multiply each term of the second polynomial by $4p$.
$= 12p^3 - 24p^2 + 8p$	Multiply the coefficients. Add the exponents on equal bases.
Stop here!	No similar terms to add.
c) $(2a + 4)(3a + 5)$	
$= 2a(3a) + 2a(5) + 4(3a) + 4(5)$	Multiply each term of the second polynomial by $2a$, and then by 4.
$= 6a^2 + 10a + 12a + 20$	Multiply the coefficients. Add the exponents on equal bases.
$= 6a^2 + (10a + 12a) + 20$	Add similar terms.
$= 6a^2 + 22a + 20$	
d) $(3x + 5)(2x - 3)$	
$= 3x(2x) + 3x(-3) + 5(2x) + 5(-3)$	Multiply each term of the second polynomial by $3x$ and then by 5.
$= 6x^2 + (-9x) + 10x + (-15)$	Multiply the coefficients. Add the exponents on equal bases.
$= 6x^2 + 1x + (-15)$	$(-9 + 10)x = 1x$
$= 6x^2 + x - 15$	Write $1x$ as x, write $+(-15)$ as -15.
e) $(4b - 3)(2b - 7)$	
$= (4b + (-3))(2b + (-7))$	Use the definition of subtraction.
$= 4b(2b) + 4b(-7) + (-3)(2b) + (-3)(-7)$	Multiply each term of the second polynomial by $4b$ and then by -3.
$= 8b^2 + (-28b) + (-6b) + 21$	Multiply the coefficients. Add the exponents on equal bases.
$= 8b^2 - 34b + 21$	$(-28b) + (-6b) = ((-28) + (-6))b = -34b$.

Exercise 5.26

Multiply:

a) $(2x + 3y)^2$

b) $(3x + 2)(4x^2 - 6x + 1)$

c) $(2p^2 + p + 3)(4p^2 + 5p - 3)$

d) $(2m - 1)^3$

Solution 5.26

a) $(2x + 3y)^2$	
$= (2x + 3y)(2x + 3y)$	Write $(2x + 3y)$ two times.

$= 2x(2x) + 2x(3y) + 3y(2x) + 3y(3y)$	Multiply each term of the second polynomial by $2x$, and then by $3y$.
$= 4x^2 + 6xy + 6xy + 9y^2$	Multiply the coefficients. Add the exponents on equal bases.
$= 4x^2 + 12xy + 9y^2$	Combine similar terms.
b) $(3x + 2)(4x^2 - 6x + 1)$	This can be done either horizontally or vertically. Both solutions are presented.
Horizontally:	
$3x(4x^2) + 3x(-6x) + 3x(1)$ $+ 2(4x^2) + 2(-6x) + 2(1)$	Multiply each term of second polynomial by $3x$, and then by 2.
$= 12x^3 - 18x^2 + 3x + 8x^2 - 12x + 2$	Multiply the coefficients. Add the exponents on equal bases.
$= 12x^3 + (-18x^2 + 8x^2) + (3x - 12x) + 2$	Group similar terms together.
$= 12x^3 - 10x^2 - 9x + 2$	Add similar terms.
Vertically:	
$4x^2 - 6x + 1$ $3x + 2$	Stack the polynomials.
$8x^2 - 12x + 2$	$2(4x^2 - 6x + 1)$
$12x^3 - 18x^2 + 3x$	$3x(4x^2 - 6x + 1)$
$12x^3 - 10x^2 - 9x + 2$	Add similar terms.
c) $(2p^2 + p + 3)(4p^2 + 5p - 3)$	You can work horizontally or vertically. The vertical form is shown here.
$2p^2 + p + 3$ $4p^2 + 5p - 3$	Stack the polynomials.
$-6p^2 - 3p - 9$	$3(2p^2 + p + 3) = -6p^2 - 3p - 9$
$10p^3 + 5p^2 + 15p$	$5p(2p^2 + p + 3) = 10p^3 + 5p^2 + 15p$
$8p^4 + 4p^3 + 12p^2$	$4p^2(2p^2 + p + 3) = 8p^4 + 4p^3 + 12p^2$
$8p^4 + 4p^3 + 11p^2 + 12p - 9$	Combine similar terms.
d) $(2m - 1)^3 = (2m - 1)(2m - 1)(2m - 1)$	Write $(2m - 1)$ three times.
$= (2m - 1)(2m - 1)(2m - 1)$	Multiply two polynomials at a time.
$= [2m(2m) + 2m(-1) + (-1)2m$ $+ (-1)(-1)](2m - 1)$	Multiply the first two polynomials.
$= (4m^2 - 2m - 2m + 1)(2m - 1)$	Multiply the coefficients. Add the exponents on equal bases.

$= (4m^2 - 4m + 1)(2m - 1)$ — You can work horizontally or vertically. This is done horizontally.

$= 4m^2(2m) + 4m^2(-1) + (-4m)(2m)$
$+ (-4m)(-1) + 1(2m) + 1(-1)$

$= 8m^3 - 4m^2 - 8m^2 + 4m + 2m - 1$ — Add similar terms.

$= 8m^3 - 12m^2 + 6m - 1$

5.6 MULTIPLYING USING FOIL AND SPECIAL PRODUCTS

Although all multiplication of polynomials can be done with the techniques discussed in Section 5.5, several shortcuts are available to speed up your work.

MULTIPLYING USING FOIL

To multiply two binomials using FOIL:

1. Multiply the **F**irst two terms.
2. Multiply the **O**uter two terms.
3. Multiply the **I**nner two terms.
4. Multiply the **L**ast two terms.
5. Add the four terms, combining similar terms whenever possible.

Exercise 5.27

Multiply using FOIL:

a) $(y + 7)(y + 5)$
b) $(x + 4)(x - 3)$
c) $(2p - 3)(p + 6)$
d) $(3m - 1)(4m - 5)$
e) $(x - 2)(y + 4)$

Solution 5.27

a) $(y + 7)(y + 5)$

F : $y(y) = y^2$ — Multiply the first two terms.

O : $y(5) = 5y$ — Multiply the outer two terms.

I : $7(y) = 7y$ — Multiply the inner two terms.

L : $7(5) = 35$ — Multiply the last two terms.

$= y^2 + 5y + 7y + 35$ — Add the four terms.

$= y^2 + 12y + 35$ — Combine similar terms: $5y + 7y = 12y$.

b) $(x + 4)(x - 3)$

F : $x(x) = x^2$	Multiply the first two terms.
O : $x(-3) = -3x$	Multiply the outer two terms.
I : $4(x) = 4x$	Multiply the inner two terms.
L : $4(-3) = -12$	Multiply the last two terms.
$= x^2 - 3x + 4x - 12$	Add the four terms.
$= x^2 + x - 12$	Combine similar terms: $-3x + 4x = 1x = x$.

c) $(2p - 3)(p + 6)$

F : $2p(p) = 2p^2$	Multiply the first two terms.
O : $2p(6) = 12p$	Multiply the outer two terms.
I : $-3(p) = -3p$	Multiply the inner two terms.
L : $-3(6) = -18$	Multiply the last two terms.
$= 2p^2 + 12p + (-3p) + (-18)$	Add the four terms.
$= 2p^2 + 9p - 18$	Combine similar terms: $12p + (-3p) = 9p$.

d) $(3m - 1)(4m - 5)$

F : $3m(4m) = 12m^2$	Multiply the first two terms.
O : $3m(-5) = -15m$	Multiply the outer two terms.
I : $-1(4m) = -4m$	Multiply the inner two terms.
L : $-1(-5) = 5$	Multiply the last two terms.
$= 12m^2 + (-15m) + (-4m) + 5$	Add the four terms.
$= 12m^2 - 19m + 5$	Combine similar terms: $(-15m) + (-4m) = 19m$.

e) $(x - 2)(y + 4)$

F : $x(y) = xy$	Multiply the first two terms.
O : $x(4) = 4x$	Multiply the outer two terms.
I : $-2(y) = -2y$	Multiply the inner two terms.
L : $-2(4) = -8$	Multiply the last two terms.
$= xy + 4x + (-2y) + (-8)$	Add the four terms.
$= xy + 4x - 2y - 8$	There are no similar terms to combine.

SPECIAL PRODUCTS: SQUARING A BINOMIAL

To square a binomial:

1. Square the first term.
2. Multiply 2 times the first term times the last term.
3. Square the last term.
4. Add these terms together.

This shortcut is based on the fact that when the sum or difference of two terms is multiplied using FOIL, the inner and outer terms are similar terms and can be combined. Exercise 5.28 uses this shortcut and FOIL to demonstrate.

Exercise 5.28

Multiply:

a) $(4 + y)^2$

b) $(a - 3)^2$

c) $(2y + 1)^2$

d) $(2x + 3y)^2$

Solution 5.28

a) $(4 + y)^2$	Use the shortcut.
$(4)^2 = 16$	Square the first term.
$2(4)(y) = 8y$	Multiply 2 times the first term (4) times the last term (y).
$(y)^2 = y^2$	Square the last term.
$= 16 + 8y + y^2$	Add the terms together.
FOIL:	
$(4 + y)^2 = (4 + y)(4 + y)$	Use the definition of exponents.
F : $4(4) = 16$	Multiply the first two terms.
O : $4(y) = 4y$	Multiply the outer two terms.
I : $y(4) = 4y$	Multiply the inner two terms.
L : $y(y) = y^2$	Multiply the last two terms.
$= 16 + 4y + 4y + y^2$	Add the four terms together.
$= 16 + 8y + y^2$	Combine the similar terms: $4y + 4y = 8y$.

Which method do you think is easier?

b) $(a - 3)^2$	
$(a)^2 = a^2$	Square the first term.
$2(a)(-3) = -6a$	Multiply 2 times the first term (a) times the last term (–3).
$(-3)^2 = 9$	Square the last term.
$= a^2 - 6a + 9$	Add the terms together.
c) $(2y + 1)^2$	
$(2y)^2 = 4y^2$	Square the first term.
$2(2y)(1) = 4y$	Multiply 2 times the first term ($2y$) times the last term (1).

$(1)^2 = 1$	Square the last term.
$= 4y^2 + 4y + 1$	Add the terms together.
d) $(2x + 3y)^2$	
$(2x)^2 = 4x^2$	Square the first term.
$2(2x)(3y) = 12xy$	Multiply 2 times the first term $(2x)$ times the last term $(3y)$.
$(3y)^2 = 9y^2$	Square the last term.
$= 4x^2 + 12xy + 9y^2$	Add the terms together.

Note that Exercise (d) was worked in Section 5.5 using the distributive property. Compare the two techniques. Which do you think is quicker?

SPECIAL PRODUCTS: MULTIPLYING THE SUM AND DIFFERENCE OF TWO TERMS

To find the product of the sum and difference of two terms:

1. Square the first term.
2. Square the last term.
3. Subtract these two terms.

This shortcut is based on the fact that when the sum and difference of two terms are multiplied using FOIL, the inner and outer terms sum to 0. Exercise 5.29 uses this shortcut and FOIL to demonstrate.

Exercise 5.29

Multiply:

a) $(x - y)(x + y)$
b) $(2a + 3)(2a - 3)$
c) $(2m - 3n)(2m + 3n)$

Solution 5.29

a) $(x - y)(x + y)$	Use the shortcut.
$(x)^2 = x^2$	Square the first term.
$(y)^2 = y^2$	Square the last term.
$= x^2 - y^2$	Subtract.
FOIL:	
F : $x(x) = x^2$	Multiply the first two terms.
O : $x(y) = xy$	Multiply the outer two terms.
I : $-y(x) = -xy$	Multiply the inner two terms.
L : $-y(y) = -y^2$	Multiply the last two terms.

$= x^2 + xy - xy - y^2$ Add the terms.

$= x^2 - y^2$ Combine similar terms: $xy - xy = 0$.

b) $(2a + 3)(2a - 3)$

$(2a)^2 = 4a^2$ Square the first term.

$(3)^2 = 9$ Square the last term.

$= 4a^2 - 9$ Subtract.

c) $(2m - 3n)(2m + 3n)$

$(2m)^2 = 4m^2$ Square the first term.

$(3n)^2 = 9n^2$ Square the last term.

$= 4m^2 - 9n^2$ Subtract.

5.7 DIVIDING POLYNOMIALS

This section presents the language and some of the techniques used to divide a polynomial by a polynomial. Division is discussed for a monomial divisor and then for a polynomial divisor.

DESCENDING ORDER

To divide polynomials, write the terms in descending order:

1. Write the term with the highest exponent first.
2. Write the term with the next highest exponent.
3. Continue until all terms are written.

Exercise 5.30

Write the terms in descending order:

a) $-15x - 5x^2 + 9 + 2x^3$

b) $100 - 25p^5 + 20p^2 - 75p^3$

c) $9 - 15x + 2x^3 - 5x^2$

Solution 5.30

a) $-15x - 5x^2 + 9 + 2x^3 = 2x^3 - 5x^2 - 15x + 9$

Note that $9 = 9 \cdot x^0$ and has the lowest exponent.

The constant term will always come last when writing a polynomial in descending order.

b) $100 - 25p^5 + 20p^2 - 75p^3 = -25p^5 - 75p^3 + 20p^2 + 100$

c) $9 - 15x + 2x^3 - 5x^2 = 2x^3 - 5x^2 - 15x + 9$

DIVIDING A POLYNOMIAL BY A MONOMIAL

To divide a polynomial by a monomial:

1. Write the polynomial in descending order, if necessary.
2. Divide each term by the monomial.
   a. Divide the coefficients. Leave as fractions if the numbers do not divide evenly.
   b. Subtract the exponents on equal bases. Write the answer with positive exponents.

Exercise 5.31

Divide:

a) $\dfrac{16x^4 - 8x^3 + 2x}{2x}$

b) $(100 - 25p^5 + 20p^2 - 75p^3)$ divided by $5p^2$

c) $\dfrac{12a^6 - 8a^4 + 7a}{2a^2}$

Solution 5.31

a) $\dfrac{16x^4 - 8x^3 + 2x}{2x}$ The polynomial is already in descending order.

$= \dfrac{16x^4}{2x} - \dfrac{8x^3}{2x} + \dfrac{2x}{2x}$ Divide each term by $2x$.

$= 8x^3 - 4x^2 + 1$ Divide the coefficients; subtract the exponents.

b) $(100 - 25p^5 + 20p^2 - 75p^3)$ divided by $5p^2$

$\dfrac{-25p^5 - 75p^3 + 20p^2 + 100}{5p^2}$ Write polynomial in descending order.

$= \dfrac{-25p^5}{5p^2} - \dfrac{75p^3}{5p^2} + \dfrac{20p^2}{5p^2} + \dfrac{100}{5p^2}$ Divide each term by $5p^2$.

$= -5p^3 - 15p + 4 + \dfrac{20}{p^2}$ $\dfrac{20}{p^2} = 20p^{-2}$, but we write the answer with a positive exponent.

c) $\dfrac{12a^6 - 8a^4 + 7a}{2a^2}$

$= \dfrac{12a^6}{2a^2} - \dfrac{8a^4}{2a^2} + \dfrac{7a}{2a^2}$ Divide each term by $2a^2$.

$= 6a^4 - 4a^2 + \dfrac{7}{2a}$ Leave $\dfrac{7}{2a}$, because 2 doesn't go evenly into 7.

DIVIDING POLYNOMIALS USING LONG DIVISION

To divide 265 by 8, we use the following setup:

$$\begin{array}{r} 33 \\ 8\overline{)265} \\ \underline{24} \\ 25 \\ \underline{24} \\ 1 \end{array}$$

***Remember*:** 8 is called the **divisor**, 265 is the **dividend**, 33 is the **quotient**, and 1 is the **remainder**. Using this language, the following section provides some rules for dividing a polynomial by a polynomial.

DIVIDING A POLYNOMIAL BY A POLYNOMIAL

1. Arrange the dividend and the divisor in descending order. Insert 0s for missing powers.
2. Divide the first term of the dividend by the first term of the divisor. Write this term of the quotient on the line above the dividend.
3. Multiply the quotient times the divisor. Write this product below the dividend. Subtract and bring down the next term.
4. This term becomes the new dividend. Go back to Step 2 using the new dividend.
5. Continue the process until the degree of the remainder is less than the degree of the divisor.
6. Check by multiplying the divisor times the quotient and adding the remainder. This product should equal the dividend.

Exercise 5.32

Perform the following divisions:

a) $(2x^2 - 7x + 6) \div (x - 2)$

b) $(9 - 15x + 2x^3 - 5x^2) \div (2x + 1)$

c) $\left(x^3-5x+2\right)\div(x+3)$

Solution 5.32

a) $x-2\overline{\smash{)}2x^2-7x+6}$

Step 1: The dividend and the divisor are already in descending order. There are no missing powers.

$$\begin{array}{r} 2x \\ x-2\overline{\smash{)}2x^2-7x+6} \\ \underline{-2x^2+4x} \\ -3x+6 \end{array}$$

Step 2: $\dfrac{2x^2}{x}=2x\cdot$

Step 3: Multiply $2x(x-2)=2x^2-4x$ and write this answer below the dividend. Subtract by changing the signs in $2x^2-4x$ to $-2x^2+4x$. Bring down +6.

Step 4: $-3x+6$ is the new dividend.

$$\begin{array}{r} 2x-3 \\ x-2\overline{\smash{)}2x^2-7x+6} \\ \underline{-2x^2+4x} \\ -3x+6 \\ \underline{3x-6} \\ 0 \end{array}$$

Step 2: $\dfrac{-3x}{x}=-3$

Step 3: Multiply $-3(x-2)=-3x+6$ and write this answer below the dividend. Subtract.

Step 5: The remainder is 0.

Check:

$(x-2)(2x-3)=2x^2-3x-4x+6=2x^2-7x+6$, which is the dividend, so our work checks.

b) $2x+1\overline{\smash{)}2x^3-5x^2-15x+9}$

Step 1: Write dividend and divisor in descending order. There are no missing powers.

$$\begin{array}{r} x^2 \\ 2x+1\overline{\smash{)}2x^3-5x^2-15x+9} \\ \underline{-2x^3-x^2} \\ -6x^2-15x \end{array}$$

Step 2: $\dfrac{2x^3}{2x}=x^2$

Step 3: Multiply $x^2(2x + 1) = 2x^3 + x^2$ and write this answer below the dividend. Subtract by changing the signs of $2x^3 + x^2$ to $-2x^3 - x^2$. Bring down $-15x$.

$$\begin{array}{r} x^2 - 3x \\ 2x+1\overline{)2x^3 - 5x^2 - 15x + 9} \\ \underline{-2x^3 - x^2} \\ -6x^2 - 15x \\ \underline{6x^2 + 3x} \\ -12x + 9 \end{array}$$

Step 2: $\dfrac{-6x^2}{2x} = -3x$

Step 3: Multiply $-3x(2x + 1) = -6x^2 - 3x$ and write this answer below the dividend. Subtract and bring down +9.

$$\begin{array}{r} x^2 - 3x - 6 \\ 2x+1\overline{)2x^3 - 5x^2 - 15x + 9} \\ \underline{-2x^3 - x^2} \\ -6x^2 - 15x \\ \underline{6x^2 + 3x} \\ -12x + 9 \\ \underline{12x + 6} \\ 15 \end{array}$$

Step 2: $\dfrac{-12x}{2x} = -6$

Step 3: Multiply $-6(2x + 1) = -12x - 6$ and write this answer below the dividend. Subtract.

Step 4: The remainder = 15. The degree of $15 = 15x^0$ is 0, and the degree of the divisor is $1(2x = 2x^1)$. Division stops when the degree of the remainder is less than the degree of the divisor.

Check:

Multiply the divisor times the quotient.

$(2x + 1)(x^2 - 3x - 6) = 2x^3 - 6x^2 - 12x + x^2 - 3x - 6$

$= 2x^3 - 5x^2 - 15x - 6$

Add the remainder to the product of the divisor and the quotient.

$(2x^3 - 5x^2 - 15x - 6) + 15 = 2x^3 - 5x^2 - 15x + 9$

Note that this sum equals the original dividend, so our work checks.

c) $x+3\overline{)x^3+0x^2-5x+2}$

Step 1: Note that there is no x^2 power in the dividend. Insert $0x^2$ for this missing power.

$$\begin{array}{r} x^2 \\ x+3\overline{)x^3+0x^2-5x+2} \\ \underline{-x^3-3x^2} \\ -3x^2-5x \end{array}$$

Step 2: $\dfrac{x^3}{x}=x^2$

Step 3: Multiply $x^2(x+3)=x^3+3x^2$. Subtract by changing the signs of x^3+3x^2 to $-x^3-3x^2$. Bring down $-5x$.

$$\begin{array}{r} x^2-3x \\ x+3\overline{)x^3+0x^2-5x+2} \\ \underline{-x^3-3x^2} \\ -3x^2-5x \\ \underline{+3x^2+9x} \\ 4x+2 \end{array}$$

Step 2: $\dfrac{-3x^2}{x}=-3x$

Step 3: $-3x(x+3)=-3x^2-9x$. Write this below the dividend. Subtract by changing the signs of $-3x^2-9x$ to $3x^2+9x$. Bring down 2.

$$\begin{array}{r} x^2-3x+4 \\ x+3\overline{)x^3+0x^2-5x+2} \\ \underline{-x^3-3x^2} \\ -3x^2-5x \\ \underline{+3x^2+9x} \\ 4x+2 \\ \underline{-4x-12} \\ -10 \end{array}$$

Step 2: $\dfrac{4x}{x}=4$

Step 3: $4(x+3)=4x+12$. Write this below the dividend. Subtract.

Step 4: The degree of $-10=-10x^0$ is 0. The degree of divisor is 1. Thus, -10 is the remainder.

Check:

Multiply the divisor times the quotient and add the remainder.

$(x + 3)(x^2 - 3x + 4) + (-10)$
$= (x^3 - 3x^2 + 4x + 3x^2 - 9x + 12) + (-10)$
$= (x^3 - 5x + 12) + (-10)$
$= x^3 - 5x + 2$

This is the original dividend, so the work checks.

TEST YOURSELF

1) Simplify each of the following expressions:

(a) 4^2

(b) -5^3

(c) $(-2)^3(-2)^7$

(d) $a^6 \cdot a^4 \cdot a^2$

(e) $3^{6x} \cdot 3^{4x}$

2) Simplify each of the following expressions. Leave all answers in exponential form:

(a) $(2^4)^3$

(b) $(x^3)^4$

(c) $(4y)^3$

(d) $(-5s^2t^4)^3$

(e) $[(7a^2)]^2$

3) Simplify each of the following expressions:

(a) $\left(\frac{2}{3}\right)^5$

(b) $\left(\frac{x}{y}\right)^3$

(c) $\left(\frac{7s}{5t}\right)^2$

(d) $\left(\frac{6m \cdot 3n}{p^4}\right)^2$

(e) $\left(\frac{8a}{3b^2}\right)^3$

4) Simplify each of the following expressions:

(a) 7^0

(b) $(-4x^2)^0$

(c) $3y^0$

(d) 2^{-5}

(e) $(6a)^{-2}$

5) Simplify each of the following expressions:

(a) $\dfrac{2^6}{2^3}$

(b) $\dfrac{3^7}{3^2}$

(c) $\left(\dfrac{5x^{-4}}{x^2}\right)^3$

(d) $\left(\dfrac{3^{-2}a^{-1}b^5}{3a^0b^2}\right)^{-2}$

(e) $y^3 \cdot y^{-5} \cdot y^3$

(f) $(4x^3)^{-2}$

(g) $\dfrac{(5a^2)^3(3a^{-2})^2}{25a^4}$

(h) $\dfrac{(3x^0y^2)^{-1}}{(6x^2y)^{-2}}$

6) Write each number in scientific notation:

(a) 628,000

(b) 12

(c) 0.0047230

(d) 0.3000089

7) Write each number in expanded form:

(a) 5.89×10^3

(b) 11.92×10^6

(c) 9.797×10^{-4}

(d) 1.3×10^{-8}

8) Find the indicated product or quotients. Write the answers in scientific notation:

(a) $\left(9 \times 10^3\right) \times \left(3 \times 10^2\right)$

(b) $\left(-3 \times 10^7\right) \times \left(8 \times 10^{-5}\right)$

(c) $\left(\dfrac{3.6 \times 10^{-2}}{12 \times 10^4}\right)$

(d) $\left(\dfrac{-4.8 \times 10^{-7}}{4 \times 10^{-3}}\right)$

(e) $\left(\dfrac{72 \times 10^8}{6 \times 10^{-9}}\right)$

(f) $\dfrac{(5 \times 10^4)(8 \times 10^{-6})}{4 \times 10^2}$

(g) $\dfrac{(6 \times 10^{-5})(9 \times 10^{-8})}{2 \times 10^{-6}}$

9) Find the degree of each polynomial. State whether the polynomial is a monomial, binomial, or trinomial:

(a) 3

(b) $2abc$

(c) $9x^2 + 3xy + 4y^3$

(d) $4x^2 - 9$

(e) $x^3 - y^3$

10) Perform the indicated operations:

(a) $(3x^2 + 6x + 4) + (5x^2 + 8x + 5)$

(b) $(-2x^3 + 7x - 1) + (x^3 - 4x^2 - 4x)$

(c) $(2a^2 - a + 6) - (4a^2 + 5a + 8)$

(d) $(-2a^4 - 7a^2 + 2) - (5a^4 - 3a^3 + 8)$

11) Perform the indicated operations:

(a) Subtract $2x - 7$ from $4x - 5$.

(b) Subtract $2x^2 - 6x + 2$ from $5x^2 + 3x - 8$.

(c) Subtract $3x + 6$ from the sum of $7x + 2$ and $8x - 5$.

(d) Subtract $x^2 - 3x + 9$ from the sum of $2x^2 + 4x + 5$ and $3x^2 - 6x - 2$.

12) Multiply:

(a) $(2a^2b)(4a^3)(5b^4)$

(b) $9x^2(10x^{-8})$

(c) $(7a + 9)(6a - 3)$

(d) $4x(5x + 21)$

(e) $(8x + 9y)^2$

(f) $(3a^2 + 2a + 1)(5a^2 + 3a + 1)$

(g) $(2x - 5)^2$

(h) $(3y - 4)(3y + 4)$

(i) $(5a + 3b)(5a - 3b)$

(j) $(2x + 3y)^2$

13) Perform the following divisions:

(a) $\dfrac{15x^2 - 15x + 9}{3}$

(b) $\dfrac{100x^4 - 75x^3 + 50x^2 - 25x}{25x}$

(c) $\dfrac{40x^8 + 48x^4 - 16x^2}{8x^2}$

(d) $\dfrac{35a^7 - 40a^4 + 30a^3}{-5a^2}$

(e) $(6x^2 + 17x - 45) \div (2x + 9)$

(f) $(6x^3 + 16x^2 + 9x + 24) \div (8 + 3x)$

(g) $(x^3 + 8) \div (x + 2)$

(h) $(3 + 3y^2 - 12y + 5y^5 - 20y^4) \div (y - 4)$

TEST YOURSELF ANSWERS

1) (a) 16
   (b) -125
   (c) 1024
   (d) $(a)^{12}$
   (e) $(3)^{10x}$

2) (a) 2^{12}
   (b) x^{12}
   (c) 4^3
   (d) 5^3
   (e) 7^4a^4

3) (a) $\frac{32}{243}$
   (b) $\frac{x^3}{y^3}$
   (c) $\frac{49s^2}{25t^2}$
   (d) $\frac{324m^2n^2}{p^8}$
   (e) $\frac{512a^3}{27b^6}$

4) (a) 1
   (b) 1
   (c) 3
   (d) $\frac{45}{a^2}$
   (e) $\frac{1}{36a^2}$

5) (a) 8
   (b) 3^5
   (c) $125x^{-18}$
   (d) $\frac{3^6 a^2}{b^6}$
   (e) $\frac{1}{y^6}$
   (f) $\frac{1}{16x^6}$
   (g) $\frac{45}{a^2}$
   (h) $12x^4$

6) (a) 6.28×10^5
   (b) 1.2×10^1
   (c) 4.7230×10^{-3}
   (d) 3.000089×10^{-1}

7) (a) 5,890
   (b) 11,920,000
   (c) 0.0009797
   (d) 0.000000013

8) (a) 2.7×10^6
   (b) -2.4×10^3
   (c) 3×10^{-7}
   (d) -1.2×10^{-4}
   (e) 1.2×10^{18}
   (f) 1×10^{-3}
   (g) 2.7×10^{-6}

9) (a) 0, monomial
(b) 3, monomial
(c) 3, trinomial
(d) 2, binomial
(e) 3, binomial

10) (a) $8x^2 + 14x + 9$
(b) $-x^3 - 4x^2 + 3x - 1$
(c) $-2a^2 - 6a - 2$
(d) $-7a^4 + 3a^3 - 7a^2 - 6$

11) (a) $2x + 2$
(b) $3x^2 + 9x - 10$
(c) $12x - 9$
(d) $4x^2 + x - 6$

12) (a) $40a^5b^5$
(b) $90x^{-6}$
(c) $42a^2 + 33a - 27$
(d) $20x^2 + 84x$
(e) $64x^2 + 144xy + 81y^2$
(f) $15a^4 + 19a^3 + 14a^2 + 5a + 1$
(g) $4x^2 - 20x + 25$
(h) $9y^2 - 16$
(i) $25a^2 - 9b^2$
(j) $4x^2 + 12xy + 9y^2$

13) (a) $5x^2 - 5x + 3$
(b) $4x^3 - 3x^2 + 2x - 1$
(c) $5x^6 + 8x^2 - 2$
(d) $-7a^5 + 8a^2 - 6a$
(e) $3x - 5$
(f) $2x^2 + 3$
(g) $x^2 - 2x + 4$
(h) $5y^4 + 3y + \frac{3}{y-4}$

CHAPTER 6

Factoring

In your past you have used factoring to help you reduce fractions, for example, $\frac{4}{6}=\frac{2\cdot 2}{2\cdot 3}=\frac{2}{3}$. In this chapter you will learn several techniques to factor polynomials, how to recognize some special types of factoring problems, and how to solve factorable equations.

6.1 GREATEST COMMON FACTOR AND FACTORING BY GROUPING

The first step in any factoring problem is to factor out the greatest common factor as outlined in the following material. Factoring by grouping provides a technique for factoring polynomials with four terms and is also presented in this section.

FINDING THE GREATEST COMMON FACTOR BY TRIAL AND ERROR

The greatest common factor (GCF) of a list of terms is the largest factor (numerical coefficient and variables) that evenly divides into each term in the list. Often, you will be able to find the GCF by trial and error, as follows:

1. Write the largest number that evenly divides into each coefficient in the list.
2. Write each common variable (variables that appear in every term) to the lowest exponent that appears.
3. Multiply your answers from Step 1 and Step 2 together.

Exercise 6.1

Find the GCF of $4x^2y$, $32x^3y^2z$, and $8x^4$.

Solution 6.1

Looking at 4, 32, and 8, you can see that 4 evenly divides into each of these. x is the only variable that appears in all three terms in the list. The *lowest* exponent that appears on x is 2. So the GCF is $4x^2$.

Exercise 6.2

Find the GCF of $15y^3z^2$, $25yz^3$, and $30y^4z^4$.

Solution 6.2

Looking at 15, 25, and 30, you can see that 5 evenly divides into each of these. y and z appear in all three terms in the list. The *lowest* exponent that appears on y is 1. The *lowest* exponent that appears on z is 2. So the GCF is $5y^1z^2 = 5yz^2$.

FINDING THE GREATEST COMMON FACTOR BY FACTORING

The following are some useful rules when trial and error either doesn't appeal to you or isn't convenient to use:

1. Prime factor each numerical coefficient and rewrite the terms using exponents.
2. Underline common bases (numbers and letters that appear in *every* term).
3. The GCF is the product of the underlined bases raised to the lowest exponent that appears in any of the terms.

Exercise 6.3

Find the GCF of $4x^6y^4$ and $8x^3y^5z$.

Solution 6.3

$4x^6y^4 = 2^2x^6y^4$ $\quad 4 = 2 \cdot 2 = 2^2$ $\quad 8x^3y^5z = 2^3x^3y^5z$ $\quad 8 = 2 \cdot 2 \cdot 2 = 2^3$

$\begin{cases} \underline{2}^2\,\underline{x}^6\,\underline{y}^4 \\ \underline{2}^3\,\underline{x}^3\,\underline{y}^5 z \end{cases}$ Underline the common bases: 2, x, and y.

GCF $= 2^2x^3y^4$ The lowest exponent on 2 is 2, the lowest exponent on x is 3, and the lowest exponent on y is 4.

Exercise 6.4

Find the GCF of 72, 48, and 36.

Solution 6.4

This exercise has no variables, so you have to work only with numbers.

$72 = 8 \cdot 9 = 2^3 \cdot 3^2$
$48 = 16 \cdot 3 = 2^4 \cdot 3^1$ — Write each term using exponents.
$36 = 4 \cdot 9 = 2^2 \cdot 3^2$

$$\begin{cases} \underline{2}^3 \cdot \underline{3}^2 \\ \underline{2}^4 \cdot \underline{3}^1 \\ \underline{2}^2 \cdot \underline{3}^2 \end{cases}$$

Underline the common bases: 2 and 3.

$\text{GCF} = 2^2 \cdot 3^1 = 12$ — Write the common bases to the lowest exponent that appears.

Exercise 6.5

Find the GCF of $5x^2$, $7yz$, and $12p$.

Solution 6.5

$5x^2 = 5x^2$
$7yz = 7yz$ — Write each term using exponents
$12p = 2^2 \cdot 3p$

$$\begin{cases} 5x^2 \\ 7yz \\ 2^2 \cdot 3p \end{cases}$$

There are *no* common bases!

GCF = 1, because 1 is the only number that evenly divides 5, 7, and 12, and there are no common variables.

FACTORING OUT THE GREATEST COMMON FACTOR (GCF)

In Chapter 5, you multiplied polynomials using the distributive property, FOIL, and several shortcuts. In this chapter, you need to reverse the multiplication process, which is called **factoring.** The first step in any factoring problem is to factor out the greatest common factor. To do that follow these steps:

1. Write the GCF of all the terms followed by a set of parentheses. Leave room inside the parentheses.
2. Divide the GCF into *each* term and write that quotient inside the parentheses.

Exercise 6.6

Factor out the greatest common factor (GCF):

a) $6x^2 + 36x$
b) $20x^5 + 10x^3 + 30x^2$

c) $36r^6s^2 - 16r^3s^5$

d) $72a^3b^4 - 48a^2b^3 + 36ab^2$

e) $x(x + 3) + 2(x + 3)$

f) $2x(x - 1) - 3(x - 1)$

Solution 6.6

a) $6x^2 + 36x$

$= 6x(\quad)$ — If you can see that $6x$ evenly divides each term, skip the step in which you write each term using exponents.

$= 6x(x + 6)$ — $\frac{6x^2}{6x} = x \quad \frac{36x}{6x} = 6$

b) $20x^5 + 10x^3 + 30x^2$

$= 2^2 \cdot 5x^5 + 2 \cdot 5x^3 + 2 \cdot 3 \cdot 5x^2$ — Write each term using exponents.

$= 2 \cdot 5x^2(\quad)$ — Write the GCF, followed by parentheses.

$= 10x^2(2x^3 + x + 3)$ — $\frac{20x^5}{10x^2} = 2x^3;\ \frac{10x^3}{10x^2} = x;\ \frac{30x^2}{10x^2} = 3$

$= 10x^2(2x^3 + x + 3)$

If you can see that 10 evenly divides 20, 10, and 30, you can skip the step in which you write each term using exponents. Just remember to also factor out *common* variables to the *lowest* exponent that appears in any one term.

c) $36r^6s^2 - 16r^3s^5 = 2^2 \cdot 3^2r^6s^2 - 2^4r^3s^5$ — Write each term using exponents.

$= 2^2r^3s^2(\quad)$ — Write the GCF, followed by parentheses.

$= 4r^3s^2(9r^3 - 4s^3)$ — $\frac{36r^6s^2}{4r^3s^2} = 9r^3$ and $\frac{16r^3s^5}{4r^3s^2} = 4s^3$.

d) $72a^3b^4 - 48a^2b^3 + 36ab^2$

$= 12ab^2(\quad)$ — Write the GCF, followed by parentheses. See Exercise 6.4.

$= 12ab^2(6a^2b^2 - 4ab + 3)$ — $\frac{72a^3b^3}{12ab^2} = 6a^2b^2;\ \frac{48a^2b^3}{12ab^2} = 4ab;\ \frac{36ab^2}{12ab^2} = 3$

e) $x(x + 3) + 2(x + 3) = (x + 3)(x + 2)$ — Here, the only common factor is $(x + 3)$, and the lowest exponent that appears is an understood 1.

f) $2x(x - 1) - 3(x - 1) = (x - 1)(2x - 3)$ — The common factor is $(x - 1)$.

Note: Begin every factoring exercise by looking for the GCF.

FACTORING BY GROUPING

In this section, you will do several types of problems that involve factoring by grouping. We concentrate on problems with four terms that can be grouped as two terms plus two terms.

Here are some general rules:

1. Group together the first two terms and factor out the GCF. Group together the last two terms and factor out the GCF.
2. If the expressions in the parentheses are equal, factor out that expression as the GCF.
3. If the expressions in the parentheses are not equal, change the order of the terms and go back to Step 1.

Exercise 6.7

Factor by grouping:

a) $xa + xy + 2a + 2y$

b) $x^2 + 2x + 2x + 4$

c) $mn + 4m - 2n - 8$

d) $pq + 3q + p + 3$

e) $rs + 3t + rt + 3s$

Solution 6.7

a) $xa + xy + 2a + 2y$

$= (xa + xy) + (2a + 2y)$	Group together the first two terms and the last two terms.
$= x(a + y) + 2(a + y)$	Factor out the GCF, x, from the first two terms. Factor out the GCF, 2, from the last two terms.
$= (a + y)(x + 2)$	$a + y$ is now the GCF and can be factored out.

b) $x^2 + 2x + 2x + 4$

$= (x^2 + 2x) + (2x + 4)$	Group together the first two terms and the last two terms.
$= x(x + 2) + 2(x + 2)$	Factor out the GCF, x, from the first two terms. Factor out the GCF, 2, from the last two terms.
$= (x + 2)(x + 2)$	$x + 2$ is now the GCF and can be factored out.

c) $mn + 4m - 2n - 8$

$= (mn + 4m) + (-2n - 8)$	Group together the first two terms and the last two terms.

$= m(n + 4) - 2(n + 4)$	Factor out the GCF, m, from the first two terms. The GCF for the last two terms is 2 or –2. Using –2 as the GCF will make the terms in the second parentheses equal the terms in the first set of parentheses.
$= (n + 4)(m - 2)$	$n + 4$ is now the GCF.
d) $pq + 3q + p + 3$	
$= q(p + 3) + 1(p + 3)$	Group the first two terms together and the last two terms together, and then factor out the greatest common factors.
$= (p + 3)(q + 1)$	The common factor is now $p + 3$.
e) $rs + 3t + rt + 3s$	
$= (rs + 3t) + (rt + 3s)$	Group together the first two terms and the last two terms.
	There are no common factors in either set of parentheses. Change the order of the terms and try again.
$= rs + rt + 3s + 3t$	Change the order so that you can find a common factor.
$= (rs + rt) + (3s + 3t)$	Group together the first two terms and the last two terms.
$= r(s + t) + 3(s + t)$	Factor out the GCF, r, from the first two terms. Factor out the GCF, 3, from the last two terms.
$= (s + t)(r + 3)$	The common factor is now $s + t$.

Note in Exercise (e) that you can also reorder the terms as $rs + 3s + rt + 3t$, and you will get the same answer: $(s + t)(r + 3)$.

6.2 FACTORING TRINOMIALS OF THE FORM $ax^2 + bx + c$, WHERE $a = 1$

When the leading coefficient of a trinomial equals 1, you need to focus your attention on the factors of the constant term, c.

IDENTIFYING a, b, AND c IN A TRINOMIAL

In the trinomial $ax^2 + bx + c$, the coefficient of x^2 is a, the coefficient of x is b, and c is the constant term.

Exercise 6.8

Identify a, b, and c in each trinomial:

a) $2x^2 + 4x + 3$
b) $x^2 - 2x + 5$
c) $3x^2 - x - 4$
d) $-4x^2 + 2x - 7$
e) $5 - 3x + 7x^2$

Solution 6.8

a) $2x^2 + 4x + 3$:

$a = 2$
$b = 4$
$c = 3$

b) $x^2 - 2x + 5$:

First rewrite as $1x^2 + (-2x) + 5$

$a = 1$
$b = -2$
$c = 5$

c) $3x^2 - x - 4$:

First rewrite as $3x^2 + (-1x) + (-4)$

$a = 3$
$b = -1$
$c = -4$

d) $-4x^2 + 2x - 7$:

First rewrite as $-4x^2 + 2x + (-7)$

$a = -4$
$b = 2$
$c = -7$

e) $5 - 3x + 7x^2$:

First rewrite in descending order as $7x^2 - 3x + 5 = 7x^2 + (-3x) + 5$.

$a = 7$
$b = -3$
$c = 5$

FACTORING TRINOMIALS

When $a = 1$ in $ax^2 + bx + c$, you can factor using the following rules:

1. Identify a, b, and c.
2. List all pairs of factors of c.
3. Find a pair of factors whose sum equals b. If there is no such pair, the trinomial does not factor.
4. Write two sets of parentheses: $(x \quad)(x \quad)$.
5. Fill in the blanks with the pair from Step 2.
6. Check your answer using FOIL.

Exercise 6.9

Factor each trinomial:

a) $x^2 + 7x + 10$

b) $x^2 + x - 12$

c) $x^2 - 7x + 6$

d) $x^2 - 2x - 63$

e) $x^2 + 3x + 14$

Solution 6.9

a) In $x^2 + 7x + 10$, $a = 1$, $b = 7$, $c = 10$. — Identify *a*, *b*, and *c*.

$c = +10$	
+1	+10
+2	+ 5
−1	−10
−2	− 5

List all pairs of factors of +10.

Find a pair whose sum equals 7.

$(x\_\_\_\_)(x\_\_\_\_)$ — Set up a pair of parentheses.

$(x + 2)(x + 5)$ — Fill in the blanks.

Check:

$(x + 2)(x + 5) = x^2 + 5x + 2x + 10.$ — Check your answer using FOIL.

$= x^2 + 7x + 10$ — This equals the original polynomial, so the answer is correct.

b) In $x^2 + x - 12$, $a = 1$, $b = 1$, $c = 12$. — Identify *a*, *b*, and *c*.

$c = -12$	
+1	−12
+2	− 6
+3	+12
−2	+ 6
−3	+ 4

List all pairs of factors of −12.

Find a pair whose sum equals 1.

$(x\_\_\_\_)(x\_\_\_\_)$ — Set up a pair of parentheses.

$(x - 3)(x + 4)$ — Fill in the blanks.

Check:

$(x - 3)(x + 4) = x^2 + 4x - 3x - 12$ — Check your answer using FOIL.

$x^2 + x - 12$

c) In $x^2 - 7x + 6$, $a = 1$, $b = -7$, $c = 6$. Identify a, b, and c.

$c = +6$	
+1	+6
+2	+3
−1	−6
−2	−3

List all pairs of factors of + 6.

Find a pair whose sum equals −7.

$(x\_\_\_\_)(x\_\_\_\_)$ Set up a pair of parentheses.

$(x - 1)(x - 6)$ Fill in the blanks.

Check:

$(x - 1)(x - 6) = x^2 - 6x - x + 6$ Check your answer using FOIL.

$= x^2 - 7x + 6$

d) In $x^2 - 2x - 63$, $a = 1$, $b = -2$, $c = -63$. Identify a, b, and c.

$c = -63$	
+1	−63
+3	−21
+7	− 9
−1	+63
−3	+21
−7	+ 9

List all pairs of factors of −63.

Find a pair whose sum equals −2.

$(x\_\_\_\_)(x\_\_\_\_)$ Set up a pair of parentheses.

$(x + 7)(x - 9)$ Fill in the blanks.

Check:

$(x + 7)(x - 9) = x^2 - 9x + 7x - 63$. Check your answer using FOIL.

$= x^2 - 2x - 63$

e) In $x^2 + 3x + 14$, $a = 1$, $b = 3$, $c = 14$. Identify a, b, and c.

$c = 14$	
+1	+14
+2	+ 7
−1	−14
−2	− 7

List all pairs of factors of 14.

Find a pair whose sum equals 3.

There is no pair of factors whose sum equals 3. This trinomial does not factor. The answer may be stated as "prime" or "does not factor."

Remember: You may need to factor out the greatest common factor (GCF) before using the rules given for trinomials. Be careful to list the GCF in your final answer.

Exercise 6.10

Factor:

a) $2x^2 + 4x - 30$

b) $-3x^2 + 30x - 63$

c) $6x^3 - 18x^2 - 24x$

d) $2x^2 + 2x + 12$

Solution 6.10

a) $2x^2 + 4x - 30 = 2(x^2 + 2x - 15)$ — Factor out the GCF: 2.

In $x^2 + 2x - 15$, $a = 1$, $b = 2$, $c = -15$. — Identify a, b, and c.

$c = 15$	
+1	+15
+3	− 5
−1	+15
−3	+ 5

List all pairs of factors of −15.

Find a pair whose sum equals 2.

$= 2(x - 3)(x + 5)$ — Be careful to list the GCF in the answer.

Check:

$2(x - 3)(x + 5) = 2(x^2 + 5x - 3x - 15)$ — Use FOIL to multiply the binomials.

$= 2(x^2 + 2x - 15)$ — Combine similar terms.

$= 2x^2 + 4x - 30$ — Use the distributive property.

b) $-3x^2 + 30x - 63 = -3(x^2 - 10x + 21)$ — Factor out the GCF: −3.

In $x^2 - 10x + 21$, $a = 1$, $b = -10$, $c = 21$. — Identify a, b, and c.

$c = 21$	
+1	+21
+3	+ 7
−1	−21
−3	− 7

List all pairs of factors of 21.

Find a pair whose sum equals −10.

$= -3(x - 3)(x - 7)$ — Use the GCF in the answer.

Check:

$-3(x - 3)(x - 7)$

$= -3(x^2 - 7x - 3x + 21)$ — Use FOIL to multiply the binomials.

$= -3(x^2 - 10x + 21)$ — Combine similar terms.

$= -3x^2 + 30x - 63$ — Use the distributive property.

c) $6x^3 - 18x^2 - 24x = 6x(x^2 - 3x - 4)$	Factor out the GCF: $6x$.
In $x^2 - 3x - 4$, $a = 1$, $b = -3$, $c = -4$.	Identify a, b, and c.
$c = -4$	List all pairs of factors of -4.
$+1\ \ -4$	
$+2\ \ -2$	Find a pair whose sum equals -3.
$-1\ \ +4$	
$-2\ \ +2$	
$= 6x(x + 1)(x - 4)$	Use the GCF in the answer.

Check:

$6x(x + 1)(x - 4) = 6x(x^2 - 4x + x - 4)$	Use FOIL to multiply the binomials.
$= 6x(x^2 - 3x - 4)$	Combine similar terms.
$= 6x^3 - 18x^2 - 24x$	Use the distributive property.

d) $2x^2 + 2x + 12 = 2(x^2 + x + 6)$	Factor out the GCF: 2.
In $x^2 + x + 6$, $a = 1$, $b = 1$, $c = 6$.	Identify a, b, and c.
$c = +6$	List all pairs of factors of 6.
$+1\ \ +6$	
$+2\ \ +3$	Find a pair whose sum equals 1.
$-1\ \ -6$	
$-2\ \ -3$	

There is no pair of factors whose sum equals 1. Therefore, the only factoring that can be done is factoring out the GCF. The answer is: $2(x^2 + x + 6)$.

Exercise 6.11

Factor completely:

a) $5x^3y - 10x^2y - 75xy$

b) $3x^2 + 6xy - 24y^2$

c) $4a^3 - 20ab + 24ab^2$

d) $3a^2 - 12a - 12$

Solution 6.11

a) $5x^3y - 10x^2y - 75xy$

$5xy(x^2 - 2x - 15)$	Factor out the GCF: $5xy$.
$c = -15$. Use $-5 \cdot 3$	Find a pair of factors of -15 whose sum is -2.
$5xy(x - 5)(x + 3)$	Try the check on your own.

b) $3x^2 + 6xy - 24y^2$

$3(x^2 + 2xy - 8y^2)$	Factor out the GCF: 3.
$c = -8y^2$. Use $+4y \cdot (-2y)$.	Find a pair of factors of $-8y^2$ whose sum is $+ 2xy$.
$3(x + 4y)(x - 2y)$	

c) $4a^3 - 20ab + 24ab^2$

$4a(a^2 - 5b + 6b^2)$	Factor out the GCF: $4a$.
$c = 6b^2$. Use $-2b \cdot (-3b)$.	Find a pair of factors of $6b^2$ whose sum is $-5ab$.
$4a(a - 2b)(a - 3b)$	

d) $3a^2 - 12a - 12$

$3(a^2 - 4a - 4)$	Factor out the GCF: 3.

$c = -4$. -4 has factors of $-4 \cdot (+1)$, $-2 \cdot (+2)$, $4 \cdot (-1)$. No pair adds to $-4a$. The only factoring that can be done is factoring out the greatest common factor.

6.3 FACTORING TRINOMIALS OF THE FORM $ax^2 + bx + c$, WHERE $a \neq 1$

If you have already factored out the greatest common factor, and the coefficient of the squared term is not equal to 1, you may choose to factor by trial and error or by grouping.

FACTORING BY TRIAL AND ERROR

The following rules can be used to factor trinomials of the form $ax^2 + bx + c$, particularly when $a \neq 1$. Although the rules given in the previous section apply only when $a = 1$, the following rules can be used for any value of a. We make use of the fact that the factors of $ax^2 + bx + c$ will be a pair of binomials in which the product of the first two terms must equal ax^2, and the product of the last two terms must equal c. These steps assume that the greatest common factor has already been factored out.

To factor $ax^2 + bx + c$ by trial and error:

1. List positive pairs of factors of ax^2.
2. List pairs of factors of c.
3. List pairs of binomial factors whose first terms are factors of ax^2 and whose last terms are factors of c.

4. Eliminate combinations that have a GCF. If the original trinomial has no GCF, there cannot be a GCF in any of the factors.

5. Use FOIL to multiply the binomials. Stop when you find a pair of factors whose product equals the original trinomial.

Note: You could use positive and negative factors of ax^2, but students generally find it easier to use positive and negative factors of c. If $a < 0$, start by factoring out -1.

Exercise 6.12

Factor:

a) $2x^2 + 7x + 3$

b) $4x^2 + 7x - 2$

c) $6x^2 - 7x + 2$

d) $10x^2 - x - 3$

Solution 6.12

a) $2x^2 + 7x + 3$

$ax^2 = 2x^2$		$c = +3$	
$2x$	x	$+1$	$+3$
x	$2x$	-1	-3

List pairs of factors of ax^2 and c.

$(2x + 1)(x + 3)$

List pairs of binomial factors whose first terms are factors of $2x^2$ and whose last terms are factors of $+3$.

$(2x - 1)(x - 3)$
$(x + 1)(2x + 3)$
$(x - 1)(2x - 3)$

Use FOIL to multiply the binomials:

$$(2x + 1)(x + 3) = 2x^2 + 6x + x + 3$$
$$= 2x^2 + 7x + 3$$
$$(2x - 1)(x - 3) = 2x^2 - 6x - x + 3$$
$$= 2x^2 - 7x + 3$$
$$(x + 1)(2x + 3) = 2x^2 + 3x + 2x + 3$$
$$= 2x^2 + 5x + 3$$
$$(x - 1)(2x - 3) = 2x^2 - 3x - 2x + 3$$
$$= 2x^2 - 5x + 3$$

So $2x^2 + 7x + 3 = (2x + 1)(x + 3)$.

We could have stopped after multiplying out the first pair, because its product was equal to the original trinomial.

b) $4x^2 + 7x - 2$

$ax^2 = 4x^2$		$c = -2$	
$4x$	x	$+1$	-2
$2x$	$2x$	-1	$+2$
x	$4x$		

List pairs of factors of ax^2 and c.

$(4x + 1)(x - 2)$

List pairs of binomial factors whose first terms are factors of $4x^2$ and whose last terms are factors of -2.

$(4x - 1)(x + 2)$

$(2x + 1)(2x - 2)$

$(2x - 1)(2x + 2)$

$(x + 1)(4x - 2)$

$(x - 1)(4x + 2)$

Use FOIL to multiply the binomials:

$(4x + 1)(x - 2) = 4x^2 - 8x + x - 2$
$= 4x^2 - 7x - 2$

$(4x - 1)(x + 2) = 4x^2 + 8x - x - 2$
$= 4x^2 + 7x - 2$ This is correct.

$(2x + 1)(2x - 2)$ This can be eliminated, because $2x - 2$ has a GCF of 2.

$(2x - 1)(2x + 2)$ This can be eliminated, because $2x + 2$ has a GCF of 2.

$(x + 1)(4x - 2)$ This can be eliminated, because $4x - 2$ has a GCF of 2.

$(x - 1)(4x + 2)$ This can be eliminated, because $4x + 2$ has a GCF of 2.

Notice that we have listed $4x$ and x, as well as x and $4x$ in our list of factors for $4x^2$. By using the commutative property to form our list of factors for $4x^2$, we do not need to list -2 and $+1$, and $+2$ and -1 in our list of factors of c. If you do list factors of c in both orders, you will find duplicates in the list of pairs of binomial factors (such as $(4x - 2)(x + 1)$).

c) $6x^2 - 7x + 2$

$ax^2 = 6x^2$		$c = +2$	
$6x$	x	$+1$	$+2$
$3x$	$2x$	-1	-2
x	$6x$		
$2x$	$3x$		

List pairs of factors of ax^2 and c.

$(6x + 1)(x + 2)$

$(6x - 1)(x - 2)$

$(3x + 1)(2x + 2)$

$(3x - 1)(2x - 2)$

$(x + 1)(6x + 2)$

$(x - 1)(6x - 2)$

$(2x + 1)(3x + 2)$

$(2x - 1)(3x - 2)$

List pairs of binomial factors whose first terms are factors of $6x^2$ and whose last terms are factors of $+2$.

Use FOIL to multiply the binomials:

$(6x + 1)(x + 2) = 6x^2 + 12x + x + 2$
$= 6x^2 + 13x + 2$

$(6x - 1)(x - 2) = 6x^2 - 12x - x + 2$
$= 6x^2 - 13x + 2$

$(3x + 1)(2x + 2)$ This can be eliminated, because $2x + 2$ has a GCF of 2.

$(3x - 1)(2x - 2)$ This can be eliminated, because $2x - 2$ has a GCF of 2.

$(x + 1)(6x + 2)$ This can be eliminated, because $6x + 2$ has a GCF of 2.

$(x - 1)(6x - 2)$ This can be eliminated, because $6x - 2$ has a GCF of 2.

$(2x + 1)(3x + 2) = 6x^2 + 4x + 3x + 2$
$= 6x^2 + 7x + 2$

$(2x - 1)(3x - 2) = 6x^2 - 4x - 3x + 2$
$= 6x^2 - 7x + 2$ This is it!

So $6x^2 - 7x + 2 = (3x - 2)(2x - 1)$.

d) $10x^2 - x - 3$

$ax^2 = 10x^2$		$c = -3$	
x	$10x$	$+1$	-3
$2x$	$5x$	-1	$+3$
$10x$	x		
$5x$	$2x$		

List pairs of factors of ax^2 and c.

$(x + 1)(10x - 3)$

$(x - 1)(10x + 3)$

$(2x + 1)(5x - 3)$

$(2x - 1)(5x + 3)$

$(10x + 1)(x - 3)$

$(10x - 1)(x + 3)$

$(5x + 1)(2x - 3)$

$(5x - 1)(2x + 3)$

List pairs of binomial factors whose first terms are factors of $10x^2$ and whose last terms are factors of -3.

Use FOIL to multiply the binomials:

$(x + 1)(10x - 3) = 10x^2 - 3x + 10x - 3$

$= 10x^2 + 7x - 3$

$(x - 1)(10x + 3) = 10x^2 + 3x - 10x - 3$

$= 10x^2 - 7x - 3$

$(2x + 1)(5x - 3) = 10x^2 - 6x + 5x - 3$

$= 10x^2 - x - 3$ This is it!

So $10x^2 - x - 3 = (5x - 3)(2x + 1)$.

FACTORING $ax^2 + bx + c$ BY GROUPING

If there are many choices for factors of a and factors of c, you may prefer the following method. These steps assume that you have already factored out any common factor.

To factor $ax^2 + bx + c$ by grouping, do the following:

1. Multiply $ax^2 \cdot c$.
2. List pairs of factors of $ax^2 \cdot c$.
3. Find a pair of factors whose sum equals b.
4. Rewrite $ax^2 + bx + c$ by replacing bx with the sum of the pair of factors found in Step 3.
5. Factor this new four-term polynomial by grouping.

To help you find factors more quickly, consider the signs of b and c, assuming that a is a positive number. If b and c are both positive, the factors of c must be positive. If b is negative and c is positive, the factors of c must both be negative. If b is positive and c is negative, the larger factor of c must be positive, and the smaller factor must be negative.

If b is negative and c is negative, the larger factor must be negative, and the smaller factor must be positive. Go back and check the answers from Exercise 6.12 to verify these cases.

Sign of b	Sign of c	Sign of Factors of c	Exercise 6.12
+	+	Both must be +.	See 6.12 (a)
–	+	Both must be –.	See 6.12 (c)
+	–	The larger factor must be +; the smaller factor must be –.	See 6.12 (b)
–	–	The larger factor must be –; the smaller factor must be +.	See 6.12 (d)

Exercise 6.13

Factor:

a) $8x^2 + 18x + 9$

b) $8x^2 + 14x - 15$

c) $x^2 - 10x + 24$

d) $3x^2 - 2x - 8$

Solution 6.13

a) $8x^2 + 18x + 9$

$\underline{8x^2 \cdot 9 = 72x^2}$	Multiply $ax^2 \cdot c$.
$1x \quad 72x$	List the pairs of factors of $ax^2 \cdot c$.
$2x \quad 36x$	The sign of b is + and the sign of c is +, so both factors must be positive.

$3x \quad 24x$

$4x \quad 18x$ — Find a pair whose sum equals $18x$.

$\mathbf{6x \quad 12x}$

$8x \quad 9x$

$8x^2 + 6x + 12x + 9$ — Replace $18x$ with $6x + 12x$.

$2x(4x + 3) + 3(4x + 3)$ — Factor by grouping.

$= (4x + 3)(2x + 3)$

b) $8x^2 + 14x - 15$

$\underline{8x^2 \cdot -15 = -120x^2}$ — Multiply $ax^2 \cdot c$.

$-1x \quad 120x$ — List the pairs of factors of $ax^2 \cdot c$.

$-2x \quad 60x$ — The sign of b is + and the sign of c is –, so the larger factor must be positive and the smaller factor must be negative.

$-3x \quad 40x$

$-4x \quad 30x$

$-5x \quad 24x$

$\mathbf{-6x \quad 20x}$ — Find a pair whose sum equals $+14x$.

$-8x \quad 15x$

$-10x \quad 12x$

$8x^2 - 6x + 20x - 15$ — Replace $14x$ with $-6x + 20x$.

$2x(4x - 3) + 5(4x - 3)$ — Factor by grouping.

$= (4x - 3)(2x + 5)$

c) $x^2 - 10x + 24$

$\underline{1x^2 \cdot 24 = 24x^2}$ — Multiply $ax^2 \cdot c$.

$-1x \quad -24x$ — List pairs of factors of $ax^2 \cdot c$.

$-2x \quad -12x$ — The sign of b is – and the sign of c is +, so both factors must be negative.

$-4x \quad -6x$

$-3x \quad -8x$

$\mathbf{-4x \quad -6x}$ — Find a pair whose sum equals $-10x$.

$x^2 - 4x - 6x + 24$ — Replace $-10x$ with $-4x - 6x$.

$x(x - 4) - 6(x - 4)$ — Factor by grouping.

$= (x - 4)(x - 6)$

d) $3x^2 - 2x - 8$

$3x^2 \cdot -8 = -24x^2$		Multiply $ax^2 \cdot c$.
$1x$	$-24x$	List the pairs of factors of $ax^2 \cdot c$.
$2x$	$-12x$	The sign of b is – and the sign of c is –, so the
$3x$	$-8x$	larger factor must be negative and the smaller
$\mathbf{4x}$	$\mathbf{-6x}$	factor must be positive.
		Find a pair whose sum equals $-2x$.
$3x^2 + 4x - 6x - 8$		Replace $-2x$ with $4x - 6x$.
$x(3x + 4) - 2(3x + 4)$		Factor by grouping.
$= (3x + 4)(x - 2)$		

CHOOSING METHODS FOR FACTORING TRINOMIALS

You may find that trial and error works best for problems in which a and/or c are prime, and that grouping works well otherwise. Remember to start every factoring exercise by looking for the greatest common factor.

Exercise 6.14

Factor as completely as possible, using your method of choice. Remember to look for the GCF first:

a) $20a^3b + 15a^2b - 50ab$

b) $4x^3 + 10x^2y - 24xy^2$

c) $2s^2 - 2s - 12$

Solution 6.14

a) $20a^3b + 15a^2b - 50ab$

$= 5ab(4a^2 + 3a - 10)$		The GCF is $5ab$.
$4a^2 + 3a - 10$		
$4a^2 \cdot (-10) = -40a^2$		Neither a nor c is prime, so try factoring by grouping.
$-1a$	$40a$	
$-2a$	$20a$	List pairs of factors of $a \cdot c$.
$-4a$	$10a$	Sign of b is + and the sign of c is –, so the larger factor must be positive and the smaller factor must be negative.
$\mathbf{-5a}$	$\mathbf{8a}$	Find a pair whose sum equals $+3a$.

$4a^2 - 5a + 8a - 10$	Replace $+3a$ with $-5a + 8a$.
$a(4a - 5) + 2(4a - 5)$	Factor by grouping.
$(4a - 5)(a + 2)$	
$= 5ab(4a - 5)(a + 2)$	Remember to write the GCF in your answer.

b) $4x^3 + 10x^2y - 24xy^2$

$= 2x(2x^2 + 5xy - 12y^2)$	The GCF is $2x$.
$2x^2 + 5xy - 12y^2$	
$2x^2 \cdot (-12y^2) = -24x^2y^2$	Because a is prime, this could be done by either method.
$-1xy$ $\quad$ $24xy$ $-2xy$ $\quad$ $12xy$ $\mathbf{-3xy}$ $\quad$ $\mathbf{8xy}$	The sign of b is + and the sign of c is –, so the larger factor must be positive and the smaller factor must be negative.
$-4xy$ $\quad$ $6xy$	Find a pair whose sum equals $-5xy$.
$2x^2 - 3xy + 8xy - 12y^2$	Replace $+5xy$ with $-3xy + 8xy$.
$x(2x - 3y) + 4y(2x - 3y)$	Factor by grouping.
$(2x - 3y)(x + 4y)$	
$= 2x(2x - 3y)(x + 4y)$	Remember to write the GCF in your answer.

c) $2s^2 - 2s - 12 = 2(s^2 - s - 6)$ — The GCF is 2.

$s^2 - s - 6$	
$ax^2 = s^2$ $\quad$ $c = -6$ s $\quad s$ $\quad$ $+1$ $\ -6$ $\quad\quad\quad\quad$ $+2$ $\ -3$	Because a is prime, use trial and error. The sign of b is – and the sign of c is –, so the larger factor must be negative.
$(s + 1)(s - 6)$	List pairs of binomial factors.
$(s + 2)(s - 3)$	

Use FOIL to multiply the binomials:

$(s + 1)(s - 6) = s^2 - 6s + s - 6$	
$= s^2 - 5s - 6$	
$(s + 2)(s - 3) = s^2 - 3s + 2s - 6$	
$= s^2 - s - 6$	This is it!
$= 2(s + 2)(s - 3)$	Remember to write the GCF in your answer.

6.4 FACTORING SPECIAL POLYNOMIALS

In this section, we factor several special polynomials. The following table summarizes the formulas used in this section:

Difference of two squares	$x^2 - y^2 = (x - y)(x + y)$
Sum of two squares	$x^2 + y^2$ does not factor.
Perfect square trinomial	$x^2 + 2xy + y^2 = (x + y)^2 = (x + y)(x + y)$ $x^2 - 2xy + y^2 = (x - y)^2 = (x - y)(x - y)$
Sum of two cubes	$x^3 + y^3 = (x + y)(x^2 - xy + y^2)$
Difference of two cubes	$x^3 - y^3 = (x - y)(x^2 + xy + y^2)$

DIFFERENCE AND SUM OF TWO SQUARES

To recognize either a difference or sum of two squares, try writing each term as a perfect square. A difference of two squares has a subtraction sign between the two terms; a sum of two squares has an addition sign between the two terms.

$4x^2 - 9 = (2x)^2 - (3)^2$ Difference of two squares

$25y^2 - 1 = (5y)^2 - (1)^2$ Difference of two squares

$81p^2 - 49q^2 = (9p)^2 - (7q)^2$ Difference of two squares

$x^2 + 16 = (x)^2 + (4)^2$ Sum of two squares

$16a^2 - 7 = (4a)^2 - (7)$ *Not* a difference of two squares

To Factor a Difference of Two Squares

1. Write each term as a perfect square: $(x)^2 - (y)^2$.
2. Write two sets of parentheses: $(\_ + \_)(\_ - \_)$.
3. Use each term in parentheses from Step 1 to fill in the blanks: $(x + y)(x - y)$.
4. Check using FOIL.

Exercise 6.15

Factor each difference of two squares:

a) $4x^2 - 9$

b) $25y^2 - 1$

c) $25a^2 - 144b^2$

d) $16 - 121a^2$

Solution 6.15

a) $4x^2 - 9 = (2x)^2 - (3)^2$ — $(2x)^2 = 4x^2$. $(3)^2 = 9$.

$= (\_ + \_)(\_ - \_)$ — Write two sets of parentheses.

$= (2x + 3)(2x - 3)$ — Use the terms in parentheses from Step 1 to fill in the blanks.

Check:

$(2x + 3)(2x - 3) = 4x^2 - 6x + 6x - 9$

$= 4x^2 - 9$

b) $25y^2 - 1 = (5y)^2 - (1)^2$ — $(5y)^2 = 25y^2$.

$= (\_ + \_)(\_ - \_)$ — Write two sets of parentheses.

$= (5y + 1)(5y - 1)$ — Use the terms in parentheses from Step 1 to fill in the blanks.

Check:

$(5y + 1)(5y - 1) = 25y^2 - 5y + 5y - 1$

$= 25y^2 - 1$

c) $25a^2 - 144b^2 = (5a)^2 - (12b)^2$ — $(5a)^2 = 25a^2$. $(12b)^2 = 144b^2$.

$= (\_ + \_)(\_ - \_)$ — Write two sets of parentheses.

$= (5a + 12b)(5a - 12b)$ — Use the terms in parentheses from Step 1 to fill in the blanks.

Check:

$(5a + 12b)(5a - 12b)$

$= 25a^2 - 60ab + 60ab - 144b^2$

d) $16 - 121a^2 = (4)^2 - (11a)^2$ — $(4)^2 = 16$ $(11a)^2 = 121a^2$

$= (\_ + \_)(\_ - \_)$ — Write two sets of parentheses.

$= (4 + 11a)(4 - 11a)$ — Use the terms in parentheses from Step 1 to fill in the blanks.

Check:

Try the check on your own.

A sum of two squares, such as $x^2 + y^2$, cannot be factored. The only possible factoring would involve factoring out the GCF. (If you take more mathematics courses, you may find that a sum of two squares factors over the complex numbers or using irrational numbers, but that is beyond our needs at this point.)

To Factor a Sum of Two Squares

1. Factor out the GCF.
2. That's it!

Exercise 6.16

Factor. If a problem cannot be factored, state "cannot be factored:"

a) $x^2 + 16$

b) $4a^2 + 16$

c) $162p^2 + 98q^2$

d) $162p^2 - 98q^2$

Solution 6.16

a) $x^2 + 16$ cannot be factored.	Sum of two squares.
b) $4a^2 + 16 = 4(a^2 + 4)$	GCF, sum of two squares.
c) $162p^2 + 98q^2 = 2(81p^2 + 49q^2)$	GCF, sum of two squares.
d) $162p^2 - 98q^2 = 2(81p^2 - 49q^2)$	GCF, difference of two squares.
$81p^2 - 49q^2 = (9p)^2 - (7q)^2$	Write each term as a perfect square.
$= (9p + 7q)(9p - 7q)$	Write two sets of parentheses and fill in the perfect squares from the step above.
$= 2(9p + 7q)(9p - 7q)$	Remember to write the GCF in your final answer.

Notice the difference between (c) and (d). A sum of two squares does not factor, while a difference of two squares *does* factor!

Exercise 6.17

Factor completely:

a) $x^4 - y^4$

b) $(x - 2)^2 - 16$

c) $3x^6 - 48$

Solution 6.17

a) $x^4 - y^4 = (x^2)^2 - (y^2)^2$ — Write each term as a perfect square.

$= (x^2 + y^2)(x^2 - y^2)$ — Keep going! $x^2 - y^2$ is a difference of two squares.

$= (x^2 + y^2)[(x)^2 - (y)^2]$ — Write each term as a perfect square.

$= (x^2 + y^2)(x + y)(x - y)$ — Note that $x^2 + y^2$ is a sum of two squares and cannot be factored further.

b) $(x - 2)^2 - 16 = (x - 2)^2 - (4)^2$ — Write each term as a perfect square. $(x - 2)^2$ is already written as a perfect square.

$16 = (4)^2$.

$= (\_ + \_)(\_ - \_)$ — Write two sets of parentheses.

$= (x - 2 + 4)(x - 2 - 4)$ — Fill in the blanks with the perfect squares.

$= (x + 2)(x - 6)$ — Simplify: $-2 + 4 = 2$, $-2 - 4 = -6$.

c) $3x^6 - 48 = 3(x^6 - 16)$ — Factor out the GCF: 3.

$x^6 - 16 = (x^3)^2 - (4)^2$ — Write each term as a perfect square. To raise a power to a power, multiply the exponents, so $(x^3)^2 = x^{3 \cdot 2} = x^6$.

$= (x^3 + 4)(x^3 - 4)$ — Write two sets of parentheses and fill in the perfect squares.

$= 3(x^3 + 4)(x^3 - 4)$ — Remember to write the GCF in your final answer.

IDENTIFYING PERFECT SQUARE TRINOMIALS

A perfect square trinomial can be written as

$$(x)^2 \pm 2(x)(y) + (y)^2$$

The first and third terms are perfect squares (terms that can be written as a quantity raised to the second power). The middle term is twice the product of the terms in the parentheses. The sign stands for plus or minus and means that the middle sign could be positive or could be negative.

Exercise 6.18

Which of the following are perfect square trinomials?

a) $4a^2 + 20a + 25$

b) $25x^2 - 30xy + 9y^2$

c) $x^2 - 2xy + 4y^2$

d) $16p^2 + 8pq + 15$

Solution 6.18

a) $4a^2 + 20a + 25$

Are the first and third terms perfect squares?

$4a^2 = (2a)^2$ $25 = (5)^2$ Yes.

Is the middle term $2(2a)(5)$?

$2(2a)(5) = 20a$ Yes.

Thus, this is a perfect square trinomial.

b) $25x^2 - 30xy + 9y^2$

Are the first and third terms perfect squares?

$25x^2 = (5x)^2$ $9y^2 = (3y)^2$ Yes.

Is the middle term $2(5x)(3y)$?

$2(5x)(3y) = 30xy$ Yes.

Thus, this is a perfect square trinomial.

c) $x^2 - 2xy + 4y^2$

Are the first and third terms perfect squares?

$x^2 = (x)^2$ $4y^2 = (2y)^2$ Yes.

Is the middle term $2(x)(2y)$?

$2(x)(2y) = 4xy$ No.

Thus, this is not a perfect square trinomial.

d) $16p^2 + 8pq + 15$

Are the first and third terms perfect squares?

$16p^2 = (4p)^2$ $15 \neq (\ )^2$ No.

Thus, this is not a perfect square trinomial.

FACTORING PERFECT SQUARE TRINOMIALS

To factor a perfect square trinomial:

1. Write the trinomial as a perfect square trinomial: $(x)^2 \pm 2(x)(y) + (y)^2$.
2. Write a set of parentheses with a square outside. Use the sign of the middle term inside the parentheses: $(\_ \pm \_)^2$.
3. Fill in the blanks with terms whose perfect squares are the first and third terms: $(x)^2 \pm 2(x)(y) + (y)^2 = (x \pm y)^2$.

Exercise 6.19

Factor:

a) $4a^2 + 20a + 25$

b) $25x^2 - 30xy + 9y^2$

c) $121 - 22m + m^2$

d) $r^2 + 10rs + 9s^2$

Solution 6.19

a) $4a^2 + 20a + 25 = (2a)^2 + 2(2a)(5) + (5)^2$ — Write the trinomial as a perfect square trinomial.

$= (\_ + \_)^2$ — Write a set of parentheses with a square outside. Use a plus sign inside because the middle term is $+20a$.

$= (2a + 5)^2$ — Fill in the blanks with the terms whose perfect squares are the first and third terms.

b) $25x^2 - 30xy + 9y^2 = (5x)^2 - 2(5x)(3y) + (3y)^2$ — Write the trinomial as a perfect square trinomial.

$= (\_ - \_)^2$ — Write a set of parentheses with a square outside. Use a minus sign inside because the middle term is $-30xy$.

$= (5x - 3y)^2$ — Fill in the blanks with the terms whose perfect squares are the first and third terms.

c) $121 - 22m + m^2 = (11)^2 - 2(11)(m) + (m)^2$ — Write the trinomial as a perfect square trinomial.

$= (\_ - \_)^2$ — Write a set of parentheses with a square outside. Use a minus sign inside because the middle term is $-22m$.

$= (11 - m)^2$ — Fill in the blanks with the terms whose perfect squares are the first and third terms.

d) $r^2 + 10rs + 9s^2$ — This is not a perfect square trinomial.

Are the first and third terms perfect squares?

$r^2 = (r)^2 \quad 9s^2 = (3s)^2$ Yes.

Is the middle term $2(r)(3s)$?

$2(r)(3s) = 6rs$ No.

Try factoring by trial and error or by grouping:

$r^2 + 10rs + 9s^2 = (r + 9s)(r + s)$ Trial and error works well because $a = 1$.

FACTORING A PERFECT SQUARE TRINOMIAL AND A DIFFERENCE OF TWO SQUARES

Sometimes a problem involves both a perfect square trinomial and a difference of two squares.

Exercise 6.20

Factor completely:

a) $x^2 - 2x + 1 - b^2$

b) $(4p - q)^2 - 16$

Solution 6.20

a) $x^2 - 2x + 1 - b^2 = (x^2 - 2x + 1) - b^2$

$x^2 - 2x + 1$ is a perfect square trinomial.

$x^2 - 2x + 1 = (x)^2 - 2(x)(1) + (1)^2$ Write the trinomial as a perfect square trinomial.

$= (x - 1)^2$ Write a set of parentheses with a square outside.

Use the sign of the middle term $(-2x)$ inside the parentheses.

$= (x - 1)^2 - b^2$ This is a difference of two squares.

$= (\_ + \_)(\_ - \_)$ Write two sets of parentheses.

$= (x - 1 + b)(x - 1 - b)$ Fill in the blanks from the perfect squares $(x - 1)$ and b.

b) $(4p - q)^2 - 16 = (4p - q)^2 - (4)^2$ Write as a difference of two squares:

$= (\_ + \_)(\_ - \_)$ Write two sets of parentheses.

$= (4p - q + 4)(4p - q - 4)$ Fill in the blanks from the perfect squares: $(4p - q)$ and 4.

FACTORING THE SUM OR DIFFERENCE OF TWO CUBES

Here is a list of perfect cubes. You need to be able to quickly recognize these values.

$(1)^3 = 1$

$(2)^3 = 8$

$(3)^3 = 27$

$(4)^3 = 64$

$(5)^3 = 125$

$(6)^3 = 216$

$(7)^3 = 343$

$(8)^3 = 512$

$(9)^3 = 729$

$(10)^3 = 1{,}000$

To factor a sum or difference of two cubes:

1. Write each term as a perfect cube.
2. Use these formulas to factor:

   $(x)^3 + (y)^3 = (x + y)(x^2 - xy + y^2)$

   $(x)^3 - (y)^3 = (x - y)(x^2 + xy + y^2)$

The bad news is that you have to memorize these formulas. The good news is that there are some similarities about the two formulas that will help you:

1. Each factorization consists of a binomial times a trinomial (two terms times three terms).
2. The terms in the binomial are the cube roots (the numbers in the parentheses) when you write the original terms as perfect cubes.
3. The sign in the binomial is the same sign as in the original problem.
4. The first and third terms of the trinomial are the squares of the terms of the binomial.
5. The middle term of the trinomial is the product of the terms in the binomial, with the opposite sign of the binomial.

Exercise 6.21

Factor:

a) $8a^3 - 27$

b) $125r^3 + 216s^3$

c) $1{,}000p^6 + 1$

Solution 6.21

a) $8a^3 - 27 = (2a)^3 - (3)^3$	Rewrite each term as a perfect cube
$= (2a - 3)[(2a)^2 + (2a)(3) + (3)^2]$	The terms in the binomial are the numbers in parentheses, $2a$ and 3. The first and third terms in the trinomial are the squares of $2a$ and 3. The middle term is the product of $2a$ and 3, with the opposite sign of the binomial, so use a + sign.
$= (2a - 3)(4a^2 + 6a + 9)$	$(2a)^2 = 4a^2$, $(2a)(3) = 6a$, and $(3)^2 = 9$.
b) $125r^3 + 216s^3 = (5r)^3 + (6s)^3$	Rewrite each term as a perfect cube
$= (5r + 6s)[(5r)^2 - (5r)(6s) + (6s)^2]$	The terms in the binomial are the numbers in parentheses, $5r$ and $6s$. The first and third terms in the trinomial are the squares of $5r$ and $6s$. The middle term is the product of $5r$ and $6s$, with the opposite sign of the binomial, so use a – sign.
$= (5r + 6s)(25r^2 - 30rs + 36s^2)$	Simplify inside the parentheses.
c) $1{,}000p^6 + 1 = (10p^2)^3 + (1)^3$	Rewrite each term as a perfect cube
$= (10p^2 + 1)[(10p^2)^2 - (10p^2)(1) + (1)^2]$	The terms in the binomial are the numbers in parentheses, $10p^2$ and 1. The first and third terms in the trinomial are the squares of $10p^2$ and 1. The middle term is the product of $10p^2$ and 1, with the opposite sign of the binomial, so use a – sign:
$= (10p^2 + 1)(100p^4 - 10p^2 + 1)$	Simplify inside the parentheses.

A STRATEGY FOR FACTORING

Usually, you are not told which factoring method to apply. Practice as many factoring exercises as possible, keeping in mind that your speed will improve with experience. Here are some guidelines to help you decide which technique to apply.

1. Always factor out the greatest common factor (GCF) first, if possible.
2. If there are two terms, consider using the special formulas for difference of two squares, sum of two squares, or sum or difference of two cubes.
3. If there are three terms, consider using the formula for a perfect square trinomial, trial and error, or the $a \cdot c$ method.

4. If there are four terms, try grouping as two by two terms, 3 terms by 1 term, or 1 term by 3 terms.
5. Although you can multiply your answers as a check, if you failed to factor out the GCF, the check won't reveal this type of error.

Exercise 6.22

Factor completely:

a) $49t^2 - 64$

b) $18y^2 + 48y + 32$

c) $18xy + 27x + 30y + 45$

d) $a^3b + 4a^2b - 12ab$

e) $54x^3y - 686y^4$

f) $10x^2 + 2xy - 15y^2$

g) $144x^2y^2 + 1$

Solution 6.22

a) $49t^2 - 64$ — Two terms, both perfect squares.

$= (7t)^2 - (8)^2$ — Write each term as a perfect square.

$= (7t - 8)(7t + 8)$ — Use the formula for factoring a difference of two squares.

b) $18y^2 + 48y + 32$

$= 2(9y^2 + 24y + 16)$ — Factor out the GCF. Two terms, the first and last terms, are perfect squares. Try the formula for a perfect square trinomial.

$9y^2 + 24y + 16 = (3y)^2 + 2(3y)(4) + (4)^2$ — This fits the formula for a perfect square trinomial.

$= (3y + 4)^2$

Write a set of parentheses with a square outside, using the square roots of the first and third terms separated by the sign of the middle term.

$2(3y + 4)^2$ — Write the GCF in the answer.

c) $18xy + 27x + 30y + 45$

$= 3(6xy + 9x + 10y + 15)$ — Factor out the GCF. Because it has four terms,

$6xy + 9x + 10y + 15$ — try factoring by grouping.

$= 3x(2y + 3) + 5(2y + 3)$

$= (2y + 3)(3x + 5)$

$= 3(2y + 3)(3x + 5)$

d) $a^3b + 4a^2b - 12ab = ab(a^2 + 4a - 12)$	Factor out the GCF. For three terms, use trial and error.
$a^2 + 4a - 12 = (a + 6)(a - 2)$	Use $+6 - 2 = +4$.
$= ab(a + 6)(a - 2)$	Write the GCF in the answer.
e) $54x^3y - 686y^4 = 2y(27x^3 - 343y^3)$	Factor out the GCF. Two terms are in parentheses; both are perfect cubes.
$27x^3 - 343y^3 = (3x)^3 - (7y)^3$	Write each term as a perfect cube.
$= (3x - 7y)(9x^2 + 21xy + 49y^2)$	Use the formula for a difference of two cubes.
$2y(3x - 7y)(9x^2 + 21xy + 49y^2)$	Write the GCF in the answer.
f) $8x^2 + 2xy - 15y^2 = (4x - 5y)(2x + 3y)$	There are three terms and no GCF, so use trial and error.
g) $144x^2y^2 + 1$ cannot be factored.	This is the sum of two squares.

6.5 SOLVING EQUATIONS BY FACTORING

A quadratic equation in standard form can be written as $ax^2 + bx + c = 0$, where a, b, and c are real numbers and $a \neq 0$. Many quadratic equations can be solved by factoring and using the zero-factor property: If a and b are real numbers and $a \cdot b = 0$, then $a = 0$ or $b = 0$.

SOLVING FACTORABLE QUADRATIC EQUATIONS

To solve a factorable quadratic equation:

1. Write the quadratic equation in standard form by getting 0 alone on one side of the equal sign.
2. Factor completely.
3. Set each factor containing a variable equal to 0.
4. Solve each equation.
5. Check your answers in the original equation.

Exercise 6.23

Solve each equation:

a) $(x + 3)(x + 2) = 0$

b) $m^2 + m - 12 = 0$

c) $2p^2 = p + 3$

d) $32r^2 = 18$

e) $2y^2 = 10y$

f) $x(2x + 1) = 15$

Solution 6.23

a) $(x + 3)(x + 2) = 0$	This is already factored completely.
$x + 3 = 0$ or $x + 2 = 0$	Set each factor equal to 0.
$x = -3$ or $x = -2$	Solve each equation.
Check:	
$(-3 + 3)(-3 + 2) \stackrel{?}{=} 0$	Substitute –3 into the original equation.
$0(-1) = 0$	–3 checks.
$(-2 + 3)(-2 + 2) \stackrel{?}{=} 0$	Substitute –2 into the original equation.
$(1)0 = 0$	–2 checks.
b) $m^2 + m - 12 = 0$	The trinomial already equals 0.
$(m + 4)(m - 3) = 0$	Factor completely.
$m + 4 = 0$ or $m - 3 = 0$	Set each factor equal to 0.
$m = -4$ or $m = 3$	Solve each equation.
Check:	
$(-4)^2 + (-4) - 12 \stackrel{?}{=} 0$	Substitute –4 into original equation.
$16 - 4 - 12 = 0$	–4 checks.
$(3)^2 + (3) - 12 \stackrel{?}{=} 0$	Substitute 3 into original equation.
$9 + 3 - 12 = 0$	3 checks.
c) $2p^2 = p + 3$	Write the equation in standard form.
$2p^2 - p - 3 = 0$	Get 0 on one side of the equation.
$(2p - 3)(p + 1) = 0$	Factor completely.
$2p - 3 = 0$ or $p + 1 = 0$	Set each factor equal to 0.
$2p = 3$ or $p = -1$	Solve each equation.
$p = \frac{3}{2}$ or $p = -1$	These are proposed solutions.
Check:	
$2\left(\frac{3}{2}\right)^2 \stackrel{?}{=} \left(\frac{3}{2}\right) + 3$	Substitute $\frac{3}{2}$ into original equation.
$2\left(\frac{9}{4}\right) \stackrel{?}{=} \frac{3}{2} + \frac{6}{2}$	

$\frac{9}{2} = \frac{9}{2}$ — $\frac{3}{2}$ checks.

$2(-1)^2 \stackrel{?}{=} (-1) + 3$ — Substitute –1 into original equation.

$2(1) = 2$ — –1 checks.

d) $32r^2 = 18$ — Write the equation in standard form.

$32r^2 - 18 = 0$ — Get 0 alone on one side.

$2(16r^2 - 9) = 0$ — Find the common factor.

$2(4r - 3)(4r + 3) = 0$ — Factor the difference of two squares.

$4r - 3 = 0$ or $4r + 3 = 0$ — Note that $2 \neq 0$, so we set only factors with a variable equal to 0.

$4r = 3$ or $4r = -3$ — Solve each equation.

$r = \frac{3}{4}$ or $r = -\frac{3}{4}$ — These are proposed solutions. Try the check on your own.

e) $2y^2 = 10y$ — Write the equation in standard form.

$2y^2 - 10y = 0$ — Get 0 alone on one side.

$2y(y - 5) = 0$ — Factor completely.

$2y = 0$ or $y - 5 = 0$ — Set each factor equal to 0.

$y = 0$ or $y = 5$ — These are proposed solutions. Try the check on your own.

f) $x(2x + 1) = 15$ — Be careful! This looks factored, but 0 is not on one side.

$2x^2 + x = 15$ — Simplify the left side.

$2x^2 + x - 15 = 0$ — Get 0 alone on one side.

$(2x - 5)(x + 3) = 0$ — Factor completely.

$2x - 5 = 0$ or $x + 3 = 0$ — Set each factor equal to 0.

$2x = 5$ or $x = -3$ — Solve each equation.

$x = \frac{5}{2}$ or $x = -3$ — These are proposed solutions. Try the check on your own.

SOLVING EQUATIONS WITH MORE THAN TWO FACTORS

The zero-factor property can be used to solve equations with more than two factors. You must get zero alone on one side of the equal sign. Then set each factor containing a variable equal to 0 after the polynomial is factored.

Exercise 6.24

Solve each equation:

a) $x(x - 2)(2x + 5) = 0$

b) $(3x + 1)(x^2 - x - 6) = 0$

Solution 6.24

a) $x(x - 2)(2x + 5) = 0$ — The equation is already factored, with 0 on one side.

$x = 0$ or $x - 2 = 0$ or $2x + 5 = 0$ — Set each factor equal to 0.

$x = 0$ or $x = 2$ or $x = -\frac{5}{2}$ — Solve each equation.

Check:

$(0)((0) - 2)(2(0) + 5) \stackrel{?}{=} 0$ — Substitute 0 into the original equation.

$0(-2)(5) \stackrel{?}{=} 0$ — Simplify inside the parentheses first.

$0 = 0$ — 0 checks.

$(2)((2)-2)(2(2)+5) \stackrel{?}{=} 0$ — Substitute 2 into the original equation.

$2(0)(9) \stackrel{?}{=} 0$ — Simplify inside the parentheses first.

$0 = 0$ — 2 checks.

$\left(-\frac{5}{2}\right)\left(\left(-\frac{5}{2}\right)-2\right)\left(2\left(-\frac{5}{2}\right)+5\right) \stackrel{?}{=} 0$ — Substitute $-\frac{5}{2}$ into the original equation.

$\left(-\frac{5}{2}\right)\left(-\frac{9}{2}\right)(0) \stackrel{?}{=} 0$ — $-\frac{5}{2}-2=-\frac{5}{2}-\frac{4}{2}=-\frac{9}{2}$.

$2\left(-\frac{5}{2}\right)+5=-5+5=0$.

$0 = 0$ — $-\frac{5}{2}$ checks.

b) $(3x + 1)(x^2 - x - 6) = 0$ — This equation is not completely factored!

$(3x + 1)(x - 3)(x + 2) = 0$ — Factor the equation completely.

$3x + 1 = 0$ or $x - 3 = 0$ or $x + 2 = 0$ — Set each factor equal to 0.

$x = -\frac{1}{3}$ or $x = 3$ or $x = -2$ — Solve each equation.

Check:

Try the check on your own.

6.6 APPLICATIONS INVOLVING FACTORING

The following section contains a small sample of applications where the solution involves your factoring skills.

GEOMETRY FORMULAS

Some geometry formulas you need in this section are listed below:

Area of a rectangle	$A = LW$, where L is the length and W is the width of the rectangle.
Area of a triangle	$A = \frac{1}{2}bh$, where b is the base and h is the height of the triangle.
Pythagorean Theorem	$c^2 = a^2 + b^2$, where c is the length of the hypotenuse, and a and b are the lengths of the legs. (Note that this can be used only for right triangles.)

Exercise 6.25

The length of a rectangle is 4 feet more than the width. The area of the rectangle is 21 square feet. Find the dimensions.

Solution 6.25

Let x = the width, so $x = W$.

Then $x + 4$ = the length, so $x + 4 = L$

$x(x + 4) = 21$	$L \cdot W =$ Area.
$x^2 + 4x - 21 = 0$	Get 0 alone on one side.
$(x + 7)(x - 3) = 0$	Factor.
$x = -7$ or $x = 3$	Set each factor equal to 0 and solve.

Check:

$(-7)((-7) + 4) \stackrel{?}{=} 21$	Substitute –7 into the original equation.
$(-7)(-3) = 21$	–7 checks in the original equation.
$(3)((3)+4) \stackrel{?}{=} 21$	Substitute 3 into the original equation.
$(3)(7) = 21$	3 checks in the original equation.

Even though both answers check in the original equation, only a length of 3 feet makes sense. No length can be negative, so we cannot give –7 as an answer. Thus, the dimensions of the rectangle are 3 feet by 7 feet.

Exercise 6.26

The base of a triangle is twice the height. The area of the triangle is 16 square inches. Find the base of the triangle.

Solution 6.26

Let x = the height of the triangle, so $x = h$.

Twice the height means to multiply, so $2x$ = the base of the triangle, so $2x = b$.

$A = \frac{1}{2}bh$ This is the formula for the area of a triangle.

$16 = \frac{1}{2}(2x)(x)$ Substitute $b = 2x$ and $h = x$ into the formula.

$16 = x^2$ Simplify the right side of the equation.

$x^2 - 16 = 0$ Get 0 alone on one side.

$(x - 4)(x + 4) = 0$ Factor a difference of two squares.

$x - 4 = 0$ or $x + 4 = 0$ Set each factor equal to 0 and solve.

$x = 4$ or $x = -4$

A base cannot have a negative length, so we eliminate $x = -4$.

Check:

$16 \stackrel{?}{=} \frac{1}{2}(2(4))(4)$ Substitute 4 into the original equation.

$16 \stackrel{?}{=} \frac{1}{2}(32)$

$16 = 16$ 4 checks.

If $x = 4$, then $2x = 2(4) = 8$, so the base is 8 inches.

Exercise 6.27

The hypotenuse of a right triangle is 10 cm. One leg is 2 cm shorter than the other leg. Find the lengths of the legs.

Solution 6.27

Let x = the length of one leg, so $x = a$.

$x - 2$ = the length of the shorter leg, so $(x - 2) = b$.

$c^2 = a^2 + b^2$ — This is the Pythagorean Theorem, where c is the length of the hypotenuse and a and b are the lengths of the legs.

$(10)^2 = (x)^2 + (x - 2)^2$ — Substitute $a = x$ and $b = (x - 2)$ into the formula.

$100 = x^2 + x^2 - 4x + 4$

$100 = 2x^2 - 4x + 4$ — Add similar terms.

$0 = 2x^2 - 4x - 96$ — Get 0 alone on one side.

$2x^2 - 4x - 96 = 0$

$2(x^2 - 2x - 48) = 0$ — Factor out the GCF: 2.

$2(x - 8)(x + 6)$ — Factor $x^2 - 2x - 48 = (x - 8)(x + 6)$.

$x - 8 = 0$ or $x + 6 = 0$ — Set each factor equal to 0 and solve.

$x = 8$ or $x = -6$

A length cannot be negative, so we eliminate –6.

Check:

$10^2 \stackrel{?}{=} (8)^2 + ((8) - 2)^2$ — Substitute 8 into the original equation.

$100 \stackrel{?}{=} 64 + (6)^2$ — Simplify inside the parentheses.

$100 \stackrel{?}{=} 64 + 36100 = 100$ — 8 checks.

If $x = 8$, $x - 2 = 8 - 2 = 6$, so the legs are 8 cm and 6 cm.

INTEGER APPLICATIONS

Recall from Section 2.5 that consecutive even or odd integers can be represented by x, $x + 2$, $x + 4$ and so on.

Exercise 6.28

The product of two consecutive even integers is 14 more than 7 times their sum. Find the integers.

Solution 6.28

Let x = the first even integer.

$x + 2$ = the next consecutive even integer.

$x(x + 2) = 14 + 7(x + (x + 2))$	The product is 14 more than 7 times their sum. Write an equation.
$x^2 + 2x = 14 + 7(2x + 2)$	Simplify both sides of the equation.
$x^2 + 2x = 14 + 14x + 14$	Multiply.
$x^2 + 2x = 28 + 14x$	Combine similar terms.
$x^2 - 12x - 28 = 0$	Get 0 alone on one side.
$(x - 14)(x + 2) = 0$	Factor.
$x - 14 = 0$ or $x + 2 = 0$	Set each factor equal to 0.
$x = 14$ or $x = -2$	Solve each equation.

Check:

$(14)((14) + 2) \stackrel{?}{=} 14 + 7((14) + (14 + 2))$	Substitute 14 into the original equation.
$(14)(16) = 14 + 7(14 + 16)$	Simplify inside the parentheses.
$224 \stackrel{?}{=} 14 + 7(30)$	
$224 \stackrel{?}{=} 14 + 210$	Multiply before adding.
$224 = 224$	14 checks.
$(-2)((-2) + 2) \stackrel{?}{=} 14 + 7((-2) + ((-2) + 2))$	Substitute –2 into the original equation.
$(-2)(0) \stackrel{?}{=} 14 + 7(-2 + 0)$	Simplify inside the parentheses.
$0 \stackrel{?}{=} 14 + 7(-2)$	
$0 = 14 - 14$	–2 checks.

There are two sets of answers. If $x = 14$, $x + 2 = 16$. If $x = -2$, $x + 2 = 0$. So the integers are either 14 and 16 or –2 and 0.

TEST YOURSELF

1) Find the GCF:

(a) $32x^2y$, $4x^4$, $8x^2y$

(b) $15x^3y^2z$, $35x^4yz^2$, $40xy^3z$

(c) 13, 52, 104

(d) $9x^2y^8$, $27x^3y^7z^2$

(e) $3x^2y$, $5ab$, $7s^3t^4$

2) Factor the following by taking out the greatest common factor:

(a) $30x^5 + 10x^3 + 20x$

(b) $a^5 - 5a^4 + 2a^3$

(c) $4a^3b^3 - 20a^2b^2 - 20ab$

(d) $9a(5a + 7) - 2(5a + 7)$

(e) $3x(x - 5) + 2(x - 5)$

(f) $4xy(2x + 1) + 5(2x + 1)$

3) Factor the following by grouping:

(a) $st + 4t + 2rs + 8r$

(b) $pq + 2r + pr + 2q$

(c) $10xy + 35x + 6y + 21$

(d) $21x^2 + 24y + 28x + 18xy$

4) Factor each trinomial:

(a) $x^2 + 3x + 2$

(b) $x^2 + 7x + 12$

(c) $a^2 - 7ab + 10b^2$

(d) $m^2 + mn - 6n^2$

(e) $x^2 + 7xy + 10y^2$

5) Factor completely. Be careful to first factor out the greatest common factor (GCF):

(a) $4x^2 - 20x - 56$

(b) $3y^3 + 9y^2 - 12y$

(c) $2x^3y + 12x^2y^2 + 16xy^3$

(d) $4a^4 - 32a^3b + 60a^2b^2$

(e) $5x^4y + 15x^3y + 5x^2y$

6) Factor completely:

(a) $6x^2 - 7x - 20$

(b) $6p^2 - 13p + 5$

(c) $8y^2 + 30y - 27$

(d) $24m^2 - 114m + 135$

(e) $2x^2 + 9x + 4$

(f) $12x^2 + 25x + 12$

(g) $10x^2 - x - 3$

7) Factor completely:

(a) $15x^2 + 8y + 10x + 12xy$

(b) $6x^3 + 2x^2 - 27xy - 9y$

(c) $4x + 5y + 35xy + 28x^2$

(d) $48p^3r + 120p^3 + 18p^2qr + 45p^2q$

(e) $12ab + 3ac - 20b - 5c$

8) Factor completely:

(a) $9x^2 - 49$

(b) $49t^2 - 64$

(c) $a^2 - 28a + 196$

(d) $9y^2 + 24y + 16$

(e) $18a^2 - 12ab + 2b^2$

9) Factor completely:

(a) $x^2 + 9$

(b) $16a^2 + 25$

(c) $98s^2 + 162t^2$

(d) $169a^2 - 36b^2$

10) Factor completely:

(a) $144 - 24x + x^2$

(b) $361 + 38a + a^2$

(c) $9s^2 + 24st + 16t^2$

(d) $9u^2 - 12uv + 4v^2$

(e) $(2a + 1)^2 - 4$

11) Factor completely:

(a) $125x^3 - y^3$

(b) $8m^3n^3 + 343$

(c) $128r^3 - 250s^3$

(d) $24 + 81p^3$

12) Factor completely:

(a) $6a^3 - 39a^2 + 60a$

(b) $12a^3 - 34a^2 + 10a$

(c) $12a^3 - 40a^2 + 28a$

(d) $10a^3 - 55a^2 + 75a$

(e) $14a^3 - 34a^2 + 12a$

13) Solve each equation:

(a) $(x - 3)(2x + 1) = 0$

(b) $2x(x + 5) = 0$

(c) $a^2 - 10a + 24 = 0$

(d) $y^2 = 4y$

(e) $6b^2 + 7b = -2$

14) Solve each problem:

(a) The hypotenuse of a right triangle is 25 cm. One leg is 5 cm longer than the other. Find the lengths of both legs.

(b) The product of two integers is 54 and the sum of the numbers is 15. Find both integers.

(c) The length of a rectangle is 2 feet less than three times the width. The area of the rectangle is 65 square feet. Find the dimensions.

(d) The product of two consecutive odd integers is 255. Find the integers.

(e) The hypotenuse of a right triangle is 20 inches. One leg is 4 inches longer than the other leg. Find the lengths of the two legs.

TEST YOURSELF ANSWERS

1) (a) $4x^2$

(b) $5xyz$

(c) 13

(d) $9x^2y^7$

(e) 1

2) (a) $10x(3x^4 + x^2 + 2)$

(b) $a^3(a^2 - 5a + 2)$

(c) $4ab(a^2b^2 - 5ab - 5)$

(d) $(5a + 7)(9a - 2)$

(e) $(x - 5)(3x + 2)$

(f) $(2x + 1)(4xy + 5)$

3) (a) $(s + 4)(t + 2r)$

(b) $(p + 2)(q + r)$

(c) $(2y + 7)(5x + 3)$

(d) $(7x + 6y)(3x + 4)$

4) (a) $(x + 2)(x + 1)$

(b) $(x + 4)(x + 3)$

(c) $(a - 2b)(a - 5b)$

(d) $(m + 3n)(m - 2n)$

(e) $(x + 5y)(x + 2y)$

5) (a) $4(x - 7)(x + 2)$

(b) $3y(y + 4)(y - 1)$

(c) $2xy(x + 4y)(x + 2y)$

(d) $4a^2(a - 3b)(a - 5b)$

(e) $5x^2y(x^2 + 3x + 1)$

6) (a) $(2x - 5)(3x + 4)$

(b) $(3p - 5)(2p - 1)$

(c) $(4y - 3)(2y + 9)$

(d) $3(2m - 5)(4m - 9)$

(e) $(2x + 1)(x + 4)$

(f) $(3x + 4)(4x + 3)$

(g) $(5x - 3)(2x + 1)$

7) (a) $(5x + 4y)(3x + 2)$

(b) $(2x^2 - 9y)(3x + 1)$

(c) $(4x + 5y)(7x + 1)$

(d) $3p^2(8p + 3q)(2r + 5)$

(e) $(3a - 5)(4b + c)$

8) (a) $(3x + 7)(3x - 7)$

(b) $(7t - 8)(7t + 8)$

(c) $(a - 14)^2$

(d) $(3y + 4)^2$

(e) $2(3a - b)^2$

9) (a) Cannot be factored.

(b) Cannot be factored.

(c) $2(49s^2 + 81t^2)$

(d) $(13a + 6b)(13a - 6b)$

10) (a) $(12 - x)^2$

(b) $(19 + a)^2$

(c) $(3s + 4t)^2$

(d) $(3u - 2v)^2$

(e) $(2a + 3)(2a - 1)$

11) (a) $(5x - y)(25x^2 + 5xy + y^2)$

(b) $(2mn + 7)(4m^2n^2 - 14mn + 49)$

(c) $2(4r - 5s)(16r^2 + 20rs + 25s^2)$

(d) $3(2 + 3p)(4 - 6p + 9p^2)$

12) (a) $3a(2a - 5)(a - 4)$

(b) $2a(3a - 1)(2a - 5)$

(c) $4a(3a - 7)(a - 1)$

(d) $5a(2a - 5)(a - 3)$

(e) $2a(7a - 3)(a - 2)$

13) (a) $3, -\frac{1}{2}$

(b) 0, –5

(c) 4, 6

(d) 0, 4

(e) $-\frac{1}{2}, -\frac{2}{3}$

14) (a) 15 cm, 20 cm

(b) 6, 9

(c) 5 ft, 13 ft

(d) 15, 17 and –17, –15

(e) 12 in., 16 in.

CHAPTER 7

Rational Expressions

This chapter includes techniques for working with rational expressions such as reducing rational expressions and adding, subtracting, multiplying, and dividing rational expressions. Factoring skills are essential, so you may need to refer to chapter 6 to reinforce your factoring skills.

7.1 REDUCING RATIONAL EXPRESSIONS

In this section, we reduce rational expressions; that is, expressions in which the numerator and denominator are polynomials. Because the denominator may contain variables, we need to discuss restrictions to ensure that the denominator does not equal 0.

RESTRICTING VARIABLES IN DENOMINATORS

To find restrictions on variables in denominators, keep the following in mind:

1. Look for a variable in the denominator; there are no restrictions if the denominator does not contain a variable.
2. If the denominator does contain a variable, set the denominator equal to 0.
3. Solve the resulting equation, remembering to factor if necessary.
4. Use the solutions from Step 3 as restrictions.

Exercise 7.1

State any restrictions on the variable in the following rational expressions:

a) $\dfrac{4x-5}{2}$

b) $\dfrac{x+1}{x-4}$

c) $\dfrac{2x+1}{x^2-x-6}$

Solution 7.1

a) $\dfrac{4x-5}{2}$ With no variable in the denominator, there are no restrictions on x.

b) $\dfrac{x+1}{x-4}$; $x-4=0$ Set the denominator equal to 0.

$x = 4$ Solve.

$x \neq 4$ State the restriction.

c) $\dfrac{2x+1}{x^2-x-6}$

$x^2 - x - 6 = 0$ Set the denominator equal to 0.

$(x-3)(x+2) = 0$ Factor.

$x - 3 = 0$ or $x + 2 = 0$ Set each factor equal to 0.

$x = 3$ or $x = -2$ Solve.

$x \neq 3$ and $x \neq -2$ State the restrictions.

REDUCING RATIONAL EXPRESSIONS

In the past, you have reduced fractions like $\dfrac{6}{9} = \dfrac{2 \cdot 3}{3 \cdot 3} = \dfrac{2}{3}$, where the common factor of 3 was divided out of the numerator and denominator. Although you may do the factoring step mentally for an example like $\dfrac{6}{9}$, you will need to write down your factors for the polynomials in the numerators and denominators of rational expressions.

To reduce rational expressions, do the following:

1. Factor the numerator.
2. Factor the denominator.
3. Divide out the common factor(s).

Exercise 7.2

Reduce to the lowest terms:

a) $\dfrac{4x-8}{5x-10}$

b) $\dfrac{x^2+2x-3}{x^2+7x+12}$

c) $\dfrac{3x^2-75}{6x^2+42x+60}$

d) $\dfrac{x^2+2x-xy-2y}{px+2p+qx+2q}$

e) $\dfrac{(x^2+4)(x+2)}{x^4-16}$

Solution 7.2

a) $\dfrac{4x-8}{5x-10}$

$=\dfrac{4(x-2)}{5(x-2)}$ Factor the numerator; factor the denominator.

$=\dfrac{4(x-2)}{5(x-2)}$ Divide out the common factor: $x - 2$.

$=\dfrac{4}{5}$

b) $\dfrac{x^2+2x-3}{x^2+7x+12}$

$=\dfrac{(x-1)(x+3)}{(x+4)(x+3)}$ Factor the numerator; factor the denominator.

$$= \frac{(x-1)(\boldsymbol{x+3})}{(x+4)(\boldsymbol{x+3})}$$

Divide out the common factor: $x + 3$.

$$= \frac{x-1}{x+4}$$

c) $$\frac{3x^2-75}{6x^2+42x+60}$$

$$= \frac{3(x^2-25)}{6(x^2+7x+10)}$$

Factor out the GCF.

$$= \frac{3(x-5)(x+5)}{6(x+2)(x+5)}$$

The numerator is the difference of two squares. The denominator was factored by trial and error.

$$= \frac{\boldsymbol{3}(x-5)(\boldsymbol{x+5})}{\boldsymbol{3}\cdot 2(x+2)(\boldsymbol{x+5})}$$

Divide out the common factors: 3 and $x + 5$.

$$= \frac{x-5}{2(x+2)}$$

d) $$\frac{x^2+2x-xy-2y}{px+2p+qx+2q}$$

$$= \frac{x(x+2)-y(x+2)}{p(x+2)+q(x+2)}$$

There are four terms, so factor by grouping.

$$= \frac{(x+2)(x-y)}{(x+2)(p+q)}$$

Finish factoring by grouping before trying to divide out the common factors.

$$= \frac{(\boldsymbol{x+2})(x-y)}{(\boldsymbol{x+2})(p+q)}$$

Divide out the common factor: $x + 2$.

$$= \frac{(x-y)}{(p+q)}$$

e) $$\frac{(x^2+4)(x+2)}{x^4-16}$$

$$= \frac{(x^2+4)(x+2)}{(x^2-4)(x^2+4)}$$

Factor the denominator as a difference of two squares.

$$= \frac{(x^2+4)(x+2)}{(x-2)(x+2)(x^2+4)}$$ Factor $x^2 - 4 = (x-2)(x+2)$.

$$= \frac{1}{x-2}$$ Divide out common factors: (x^2+4) and $x+2$.

At first glance, an expression like $\frac{b-a}{a-b}$ may not look like it can be reduced. But consider the following:

$$= \frac{b-a}{a-b} = \frac{-a+b}{a-b}$$ Using the commutative property: $b - a = -a + b$.

$$= \frac{-1(a-b)}{a-b}$$ Factor out -1.

$$= \frac{-1(a-b)}{a-b}$$ Divide out the common factor: $a - b$.

$= -1$

You do not need to do all these steps if you understand and recognize these forms of -1. To recognize rational expressions that reduce to -1, do the following:

1. Make sure the factors in the numerator and denominator have subtraction signs between the terms.
2. Use the commutative property to change the order of the terms.
3. Write -1 as a factor in the numerator and divide out the common factors.

Exercise 7.3

Write in lowest terms:

a) $\frac{p-2}{2-p}$

b) $\frac{3x-3}{2-2x}$

c) $\frac{x^2-9}{3-x}$

d) $\frac{m-2n}{2n+m}$

e) $\frac{2x+y}{y+2x}$

Solution 7.3

a) $\dfrac{p-2}{2-p}=\dfrac{-2+p}{2-p}$ Using the commutative property: $p-2=-2+p$.

$=\dfrac{-1(2-p)}{2-p}$ Factor out -1.

$=-1$ Divide out the common factor: $2-p$.

b) $\dfrac{3x-3}{2-2x}=\dfrac{3(x-1)}{2(1-x)}$ Factor the numerator and the denominator.

$=\dfrac{3(-1)}{2}$ $\dfrac{x-1}{1-x}=-1$.

$=\dfrac{-3}{2}=-\dfrac{3}{2}$

c) $\dfrac{x^2-9}{3-x}=\dfrac{(x-3)(x+3)}{3-x}$ Factor the numerator.

$=-1(x+3)$ $\dfrac{x-3}{3-x}=-1$.

d) $\dfrac{m-2n}{2n+m}$ This is already in lowest terms. There is no subtraction sign in the denominator.

e) $\dfrac{2x+y}{y+2x}=\dfrac{2x+y}{2x+y}$ There are no subtraction signs, but this can be reduced.

$=1$

Warning: It is a very tempting to divide out terms instead of factors. Don't take the bait! Here are a couple of examples:

$\dfrac{2x+6}{2}\neq 2x+3$ Here, 6 is a term, not a factor.

$\dfrac{2x+6}{2}=\dfrac{2(x+3)}{2}=x+3$ Here, 2 is a factor and can be divided out.

7.2 MULTIPLYING AND DIVIDING RATIONAL EXPRESSIONS

The same rules used for multiplying and dividing rational numbers are used for multiplying and dividing rational expressions:

$$\frac{A}{B}\cdot\frac{C}{D}=\frac{AC}{BD} \qquad B, D \neq 0$$

$$\frac{A}{B}\div\frac{C}{D}=\frac{A}{B}\cdot\frac{D}{C}=\frac{AD}{BC} \qquad B, D, C \neq 0$$

We will use factoring to simplify this process; see the following example:

Exercise 7.4

Multiply or divide the expressions as indicated. Reduce all answers to the lowest terms:

a) $\dfrac{3x^2}{9x}\cdot\dfrac{18x}{15x}$

b) $\dfrac{25y^8}{15y^6}\div\dfrac{9y}{10y^4}$

c) $\dfrac{a^2-5a+6}{2a+4}\cdot\dfrac{a+2}{2a-6}$

d) $\dfrac{6x^2-54}{2x^2+3x+1}\div\dfrac{4x-12}{2x^2-9x-5}$

e) $\dfrac{x^2-6x+9}{x^2-1}\div\dfrac{3-x}{1+x}$

f) $\dfrac{2y^2+11y+15}{y^2-2y-15}\cdot\dfrac{y^2-25}{2y^2+3y-5}$

Solution 7.4

a) $\dfrac{3x^2}{9x}\cdot\dfrac{18x}{15x}$

$$=\frac{3x^2\cdot 18x}{9x\cdot 15x}$$ Multiply numerators; multiply denominators.

$$=\frac{3\cdot x\cdot x\cdot 9\cdot 2\cdot x}{9\cdot x\cdot 5\cdot 3\cdot x}$$ Factor.

$$=\frac{2x}{5}$$ Divide out the common factors: 3, 9, x, and x.

If you completed the multiplication in the numerator and denominator before factoring, your work would look like this:

$$\frac{3x^2 \cdot 18x}{9x \cdot 15x} = \frac{54x^3}{135x^2}$$

You would now have to factor:

$$\frac{2 \cdot 3 \cdot 3 \cdot 3 \cdot x \cdot x \cdot x}{3 \cdot 3 \cdot 3 \cdot 5 \cdot x \cdot x},$$

which would still reduce to $\frac{2x}{5}$. It is usually easier to factor first, rather than multiplying out the numerator and denominator, but either method will work.

b) $\frac{25y^8}{15y^6} \div \frac{9y}{10y^4}$

$= \frac{25y^8}{15y^6} \cdot \frac{10y^4}{9y}$ — Invert the divisor and multiply.

$= \frac{25y^8 \cdot 10y^4}{15y^6 \cdot 9y}$ — Multiply numerators; multiply denominators.

$= \frac{5 \cdot 5 \cdot y^6 \cdot y \cdot y \cdot 10y^4}{5 \cdot 3 \cdot y^6 \cdot 9 \cdot y}$ — Factor.

$= \frac{50y^5}{27}$ — Divide out the common factors: 5, y^6, y.

c) $\frac{a^2 - 5a + 6}{2a + 4} \cdot \frac{a + 2}{2a - 6}$

$= \frac{(a-3)(a-2)}{2(a+2)} \cdot \frac{a+2}{2(a-3)}$ — Factor.

$= \frac{(a-3)(a-2)(a+2)}{2(a+2)2(a-3)}$ — Multiply numerators; multiply denominators.

$= \frac{a-2}{4}$ — Divide out the common factors: $(a - 3)$ and $(a + 2)$.

d) $\frac{6x^2 - 54}{2x^2 + 3x + 1} \div \frac{4x - 12}{2x^2 - 9x - 5}$

$= \frac{6x^2 - 54}{2x^2 + 3x + 1} \cdot \frac{2x^2 - 9x - 5}{4x - 12}$ — Invert the divisor and multiply.

$= \dfrac{6(x^2-9)}{(2x+1)(x+1)} \cdot \dfrac{(2x+1)(x-5)}{4(x-3)}$ Factor.

$= \dfrac{2 \cdot 3(x-3)(x+3)(2x+1)(x-5)}{(2x+1)(x+1)2 \cdot 2(x-3)}$ Factor and multiply.

$= \dfrac{3(x+3)(x-5)}{2(x+1)}$ Divide out common factors: 2, $(x - 3)$, and $(2x + 1)$.

Answers are usually left in factored form.

e) $\dfrac{x^2-6x+9}{x^2-1} \div \dfrac{3-x}{1+x}$

$= \dfrac{x^2-6x+9}{x^2-1} \cdot \dfrac{1+x}{3-x}$ Invert the divisor and multiply.

$= \dfrac{(x-3)(x-3)(1+x)}{(x-1)(x+1)(3-x)}$ Factor and multiply.

$= \dfrac{-1(x-3)}{x-1}$ Divide out common factors: $\dfrac{x-3}{3-x} = -1$ and $\dfrac{1+x}{x+1} = 1$. You may find an answer like $\dfrac{-1(x-3)}{x-1}$ rewritten as $\dfrac{-1(x-3)}{x-1} = \dfrac{-x+3}{x-1}$. If so, use the distributive property.

$= \dfrac{3-x}{x-1}$ Use the commutative property: $-x + 3 = 3 - x$.

f) $\dfrac{2y^2+11y+15}{y^2-2y-15} \cdot \dfrac{y^2-25}{2y^2+3y-5}$

$= \dfrac{(2y+5)(y+3)}{(y+3)(y-5)} \cdot \dfrac{(y-5)(y+5)}{(2y+5)(y-1)}$ Factor.

$= \dfrac{y+5}{y-1}$ Divide out common factors: $(2y + 5)$, $(y + 3)$, and $(y - 5)$.

7.3 WRITING RATIONAL EXPRESSIONS USING THE LEAST COMMON MULTIPLE

To add or subtract rational expressions, you *must* have a common denominator. The least common denominator is the least common multiple of the given denominators. You may want to review Method 2 for finding the least common denominator in Chapter 1. To find the least common multiple:

1. Factor each expression, using exponents where possible.
2. Recognize that the least common multiple (LCM) is the product of each different factor raised to the highest exponent that appears in any of the given expressions.

Exercise 7.5

Find the least common multiple (LCM):

a) $12x^2y$; $8xy^3$
b) $x^2 + 6x + 9$; $4x + 12$
c) $x^2 + x - 2$; $x^2 - 6x + 5$

Solution 7.5

a) $12x^2y = 2^2 \cdot 3^1 \cdot x^2 \cdot y^1$ — Factor $12 = 4 \cdot 3 = 2^2 \cdot 3^1$.

$8xy^3 = 2^3 \cdot x^1 \cdot y^3$ — Factor $8 = 2^3$.

$\text{LCM} = 2^3 \cdot 3^1 \cdot x^2 \cdot y^3$ — The highest power of 2 is 3, so use 2^3.

The highest power of 3 is 1, so use 3^1.

The highest power of x is 2, so use x^2.

The highest power of y is 3, so use y^3.

$= 24x^2y^3$ — Multiply out the numerical coefficient and write the variables with exponents.

b) $x^2 + 6x + 9 = (x + 3)^2$ — Factor.

$4x + 12 = 4(x + 3)^1 = 2^2(x + 3)^1$ — Factor.

$\text{LCM} = 2^2(x + 3)^2$ — The highest power of 2 is 2, so use 2^2. The highest power of $(x + 3)$ is 2, so use $(x + 3)^2$.

$= 4(x + 3)^2$ — Multiply out the numerical coefficient.

c) $x^2 + x - 2 = (x + 2)(x - 1)$ — Factor.

$x^2 - 6x + 5 = (x - 5)(x - 1)$ — Factor.

LCM = $(x + 2)(x - 1)(x - 5)$

The highest power of each factor is 1. Be careful here—do not try to use $(x - 1)^2$, because there is no power of 2 on either factor of $(x - 1)$.

Leave the LCM in factored form.

FINDING THE LEAST COMMON DENOMINATOR (LCD)

The least common denominator (LCD) is the least common multiple of all the denominators. Follow the rules given for finding the least common multiple using the denominators of the rational expressions.

Exercise 7.6

Find the least common denominator (LCD):

$$\frac{4}{x^2 - x - 6}, \frac{6}{x^2 + x - 2}$$

Solution 7.6

$x^2 - x - 6 = (x - 3)(x + 2)$ — Factor each denominator.

$x^2 + x - 2 = (x + 2)(x - 1)$

LCD = $(x - 3)(x + 2)(x - 1)$ — Use each factor raised to the highest exponent.

Exercise 7.7

Find the LCD:

$$\frac{3}{x^2 - 2x + 1}, \frac{5}{x^2 - 2x - 3}, \frac{6}{x^2 - 4x + 3}$$

Solution 7.7

$x^2 - 2x + 1 = (x - 1)^2$ — Factor each denominator.

$x^2 - 2x - 3 = (x - 3)(x + 1)$

$x^2 - 4x + 3 = (x - 1)(x - 3)$

The highest power of $(x - 1)$ is 2, so use $(x - 1)^2$.

The highest power of $(x + 1)$ is 1, so use $(x + 1)$.

The highest power of $(x - 3)$ is 1, so use $(x - 3)$.

LCD = $(x - 1)^2(x + 1)(x - 3)$

WRITING RATIONAL EXPRESSIONS WITH THE LCD

Recall from our work with fractions that a fraction can be "built up" by multiplying the numerator and denominator by a common factor. We will use that same idea to write expressions with the least common denominator.

Exercise 7.8

Rewrite $\frac{4}{x^2-x-6}$ and $\frac{6}{x^2+x-2}$ with the least common denominator (LCD).

Solution 7.8

The LCD of $\frac{4}{x^2-x-6}$ and $\frac{6}{x^2+x-2}$ is $(x-3)(x+2)(x-1)$ (see Exercise 7.6). After writing the denominators in factored form, compare the given denominator to the LCD. Multiply by the factor (or factors) that are needed to make the given denominator equal the LCD. Remember that you must then multiply the numerator *and* denominator by this factor (or factors) to obtain equivalent rational expressions.

$$\frac{4}{x^2-x-6}=\frac{4}{(x-3)(x+2)}\cdot\left(\frac{x-1}{x-1}\right)$$

Multiply the numerator and denominator by $(x-1)$ to make the denominator equal the LCD.

$$=\frac{4x-4}{(x-3)(x+2)(x-1)}$$

Multiply numerators; keep the denominator in factored form.

$$\frac{6}{x^2+x-2}=\frac{6}{(x+2)(x-1)}\cdot\left(\frac{x-3}{x-3}\right)$$

Multiply the numerator and denominator by $(x-3)$ to make the denominator equal the LCD.

$$=\frac{6x-18}{(x+2)(x-1)(x-3)}$$

Multiply numerators; keep the denominator in factored form.

Exercise 7.9

Rewrite $\frac{3}{x^2-1}, \frac{5}{x^2-2x-3}$ and $\frac{6}{x^2-4x+3}$ with the LCD.

Solution 7.9

First find the LCD:

$x^2 - 1 = (x - 1)(x + 1)$ Factor each denominator.

$x^2 - 2x - 3 = (x - 3)(x + 1)$

$x^2 - 4x + 3 = (x - 3)(x - 1)$

LCD = $(x - 1)(x + 1)(x - 3)$ The highest exponent of $(x - 1)$ is 1, so use $(x - 1)$.

The highest exponent of $(x + 1)$ is 1, so use $(x + 1)$.

The highest exponent of $(x - 3)$ is 1, so use $(x - 3)$.

$$\frac{3}{(x-1)(x+1)} = \frac{3}{(x-1)(x+1)} \cdot \left(\frac{x-3}{x-3}\right)$$

Multiply the numerator and denominator by $(x - 3)$ to make the denominator equal the LCD.

$$= \frac{3x-9}{(x-1)(x+1)(x-3)}$$

Multiply the numerators; keep the denominator in factored form.

$$\frac{5}{(x-3)(x+1)} = \frac{5}{(x-3)(x+1)} \cdot \left(\frac{x-1}{x-1}\right)$$

Multiply the numerator and denominator by $(x - 1)$ to make the denominator equal the LCD.

$$= \frac{5x-5}{(x-3)(x+1)(x-1)}$$

Multiply the numerators; keep the denominator in factored form.

$$= \frac{6}{(x-3)(x-1)}$$

$$= \frac{6}{(x-3)(x-1)} \cdot \left(\frac{x+1}{x+1}\right)$$

Multiply the numerator and denominator by $(x + 1)$ to make the denominator equal the LCD.

$$= \frac{6x+6}{(x-3)(x-1)(x+1)}$$

Multiply the numerators; keep the denominator in factored form.

7.4 ADDING AND SUBTRACTING RATIONAL EXPRESSIONS

You must have a common denominator to add or subtract rational expressions. To add or subtract rational expressions, do the following. If the given denominators are equal, start with Step 4:

1. Factor each denominator.
2. Find the least common denominator (LCD).
3. Rewrite each rational expression with the least common denominator (LCD).
4. Add or subtract the numerators. If subtracting distribute the subtraction sign across the entire numerator. The LCD is the denominator.
5. Reduce the answer, if possible, by factoring and dividing out the common factors.

ADDING AND SUBTRACTING RATIONAL EXPRESSIONS WITH COMMON DENOMINATORS

To add two fractions with the same denominator, you add the numerators and keep the common denominator. The same rule applies for rational expressions. The following exercises involve rational expressions that have the same denominator:

Exercise 7.10

Find the sums or differences:

a) $\frac{7}{x}-\frac{9}{x}$

b) $\frac{4}{y+2}+\frac{2y}{y+2}$

c) $\frac{x^2}{x^2+3x-18}+\frac{x-12}{x^2+3x-18}$

Solution 7.10

a) $\frac{7}{x}-\frac{9}{x}=\frac{7-9}{x}$

The given denominators are equal. Subtract the numerators; use the LCD of x as the denominator.

$=\frac{-2}{x}$ or $-\frac{2}{x}$

The negative can be written in the numerator or in front of the entire fraction.

b) $\frac{4}{y+2}+\frac{2y}{y+2}$

$=\frac{4+2y}{y+2}$ — The given denominators are equal. Add the numerators; use the LCD of $y + 2$ as the denominator.

$=\frac{2(2+y)}{y+2}$ — Factor the numerator.

$=2$ — Divide out the common factor: $2 + y = y + 2$.

c) $\frac{x^2}{x^2+3x-18}+\frac{x-12}{x^2+3x-18}$ — The given denominators are equal. Add the numerators; use the LCD of $x^2 + 3x - 18$ as the denominator.

$=\frac{(x-3)(x+4)}{(x-3)(x+6)}$ — Factor the numerator and the denominator.

$=\frac{x+4}{x+6}$ — Divide out the common factor: $x - 3$.

ADDING AND SUBTRACTING RATIONAL EXPRESSIONS WITH UNLIKE COMMON DENOMINATORS

The following exercise involves adding and subtracting rational expressions when the denominators are not equal. Try to find the LCD by inspection, but if you cannot "see it," factor each denominator and use the previous work to find the LCD.

Exercise 7.11

Find the sums or differences:

a) $\frac{5x}{6}-\frac{2x}{3}$

b) $\frac{6}{x^2+x-2}-\frac{4}{x^2-x-6}$

c) $\frac{10}{x^2-25}-\frac{9}{x^2-x-20}$

d) $\dfrac{2}{p-4}+\dfrac{5}{4-p}$

Solution 7.11

a) $\dfrac{5x}{6}-\dfrac{2x}{3}$

$=\dfrac{5x}{2\cdot 3}-\dfrac{2x}{3}$ Factor the denominators.

$=\dfrac{5x}{2\cdot 3}-\dfrac{2x}{3}\cdot\left(\dfrac{2}{2}\right)$ The LCD $= 2 \cdot 3$.

$=\dfrac{5x}{2\cdot 3}-\dfrac{(2x)(2)}{2\cdot 3}$ Rewrite $\dfrac{2x}{3}$ with the LCD of 6.

$=\dfrac{5x-4x}{6}$ Subtract the numerators; use LCD of 6 as the denominator.

$=\dfrac{x}{6}$ Simplify the numerator.

b) $\dfrac{6}{x^2+x-2}-\dfrac{4}{x^2-x-6}$

$=\dfrac{6}{(x+2)(x-1)}-\dfrac{4}{(x-3)(x+2)}$ Factor the denominators.

$=\dfrac{6}{(x+2)(x-1)}\cdot\left(\dfrac{x-3}{x-3}\right)-\dfrac{4}{(x-3)(x+2)}\cdot\left(\dfrac{x-1}{x-1}\right)$ See LCD work from Exercise 7.8.

$=\dfrac{6x-18}{(x+2)(x-1)(x-3)}-\dfrac{4x-4}{(x-3)(x+2)(x-1)}$ Multiply the numerators. Keep the denominators in factored form.

$=\dfrac{6x-18-(4x-4)}{(x+2)(x-1)(x-3)}$ Subtract the numerators. Be careful to distribute the subtraction sign across the entire numerator!

$=\dfrac{6x-18-4x+4}{(x+2)(x-1)(x-3)}$ Combine similar terms in the numerator.

$=\dfrac{2x-14}{(x+2)(x-1)(x-3)}$

Even though the numerator factors to $2(x-7)$, there are no common factors to divide out, so the answer is usually left in the form shown.

c) $\dfrac{10}{x^2-25}-\dfrac{9}{x^2-x-20}$

$= \dfrac{10}{(x-5)(x+5)}-\dfrac{9}{(x-5)(x+4)}$

Factor the denominators. LCD = $(x-5)(x+5)(x+4)$.

$= \dfrac{10}{(x-5)(x+5)}\cdot\left(\dfrac{x+4}{x+4}\right)-\dfrac{9}{(x-5)(x+4)}\cdot\left(\dfrac{x+5}{x+5}\right)$

Rewrite each expression with the LCD.

$= \dfrac{10x+40}{(x-5)(x+5)(x+4)}-\dfrac{9x+45}{(x-5)(x+4)(x+5)}$

Multiply the numerators. Leave the denominators in factored form.

$= \dfrac{10x+40-(9x+45)}{(x-5)(x+5)(x+4)}$

Subtract the numerators; use the LCD as the denominator.

$= \dfrac{10x+40-9x-45}{(x-5)(x+5)(x+4)}$

Be careful to distribute the subtraction sign!

$= \dfrac{x-5}{(x-5)(x+5)(x+4)}$

Divide out common factor: $x-5$.

$= \dfrac{1}{(x+5)(x+4)}$

This is the reduced answer.

d) $\dfrac{2}{p-4}+\dfrac{5}{4-p}$

The denominators are negatives of each other. Use the commutative property and sign rules for fraction to rewrite $\dfrac{5}{4-p}$.

$= \dfrac{2}{p-4}+\dfrac{5}{-1(p-4)}$

$4-p=-p+4=-1(p-4)$.

$= \dfrac{2}{p-4}-\dfrac{5}{p-4}$

$+\dfrac{5}{-1(p-4)}=-\dfrac{5}{+(p-4)}=-\dfrac{5}{p-4}$.

$= \dfrac{2-5}{p-4}$

The denominators are now equal. Subtract the numerators; use the LCD of $(p-4)$ as the denominator.

$= \dfrac{-3}{p-4}$

Because $4 - p = -p + 4 = -1(p - 4)$, you can use either $p - 4$ or $4 - p$ as the LCD.

7.5 SOLVING EQUATIONS CONTAINING RATIONAL EXPRESSIONS

This section presents techniques for solving equations with rational expressions, not just simplifying rational expressions. To use these techniques, make sure the directions say "Solve," not "Simplify."

To solve equations containing rational expressions:

1. Factor each denominator as necessary to find the least common denominator (LCD).
2. Multiply both sides of the equation by the LCD.
3. Divide out common factors. There should be no fractions remaining after this step.
4. Solve the resulting equation.
   a) If the resulting equation is a first-degree equation, isolate the variable.
   b) If the resulting equation is a quadratic equation, get 0 on one side of the equation, and then factor and set each factor equal to 0.
5. Check each solution in the original equation.

Exercise 7.12

Solve each equation and check your answers:

a) $\dfrac{x}{6}+1=\dfrac{1}{3}$

b) $1+\dfrac{3}{x}=\dfrac{10}{x^2}$

c) $\dfrac{x}{x-4}=2+\dfrac{4}{x-4}$

d) $\dfrac{2}{x-2}+\dfrac{1}{x^2-4}=\dfrac{3}{x+2}$

e) $\dfrac{2x}{x-3}-\dfrac{x}{x+2}=\dfrac{-6}{x^2-x-6}$

f) $\dfrac{x+3}{x^2-3x-4}-\dfrac{2}{x^2+3x+2}=\dfrac{8}{x^2-2x-8}$

g) $\dfrac{2}{x-3}-\dfrac{3}{x+3}=\dfrac{12}{x^2-9}$

h) $\frac{r}{r-1}=\frac{2}{r^2-1}$

Solution 7.12

a) $\frac{x}{6}+1=\frac{1}{3}$

$6\left(\frac{x}{6}+1\right)=6\left(\frac{1}{3}\right)$ Multiply both sides of the equation by the LCD: 6.

$6\cdot\frac{x}{6}+6\cdot1=6\cdot\frac{1}{3}$ Distribute.

$x + 6 = 2$ Divide out the common factors.

$x = -4$

Check:

$\frac{-4}{6}+1\overset{?}{=}\frac{1}{3}$ Substitute $x = -4$.

$\frac{-4}{6}+\frac{6}{6}=\frac{1}{3}$

$\frac{2}{6}=\frac{1}{3}$ This is true, so the solution checks.

b) $1+\frac{3}{x}=\frac{10}{x^2}$

$x^2\left(1+\frac{3}{x}\right)=x^2\left(\frac{10}{x^2}\right)$ Multiply both sides of the equation by the LCD: x^2.

$x^2\cdot1+x^2\cdot\frac{3}{x}=x^2\cdot\frac{10}{x^2}$ Distribute.

$x^2 + 3x = 10$ This is a quadratic equation.

$x^2 + 3x - 10 = 0$ Get 0 on one side.

$(x + 5)(x - 2) = 0$ Factor.

$x = -5$ or $x = 2$ Set each factor equal to 0 and solve.

Check:

When $x = -5$,

$$1+\frac{3}{(-5)} \stackrel{?}{=} \frac{10}{(-5)^2}$$

$$1-\frac{3}{5} \stackrel{?}{=} \frac{10}{25}$$

$$\frac{2}{5} = \frac{2}{5}$$

When $x = 2$,

$$1+\frac{3}{(2)} \stackrel{?}{=} \frac{10}{(2)^2}$$

$$1+\frac{3}{2} \stackrel{?}{=} \frac{10}{4}$$

$$\frac{5}{2} = \frac{5}{2}$$

Both solutions check.

c) $\dfrac{x}{x-4} = 2+\dfrac{4}{x-4}$

$$(x-4)\frac{x}{x-4} = (x-4)\left(2+\frac{4}{x-4}\right)$$

Multiply by the LCD: $(x-4)$.

$$(x-4)\frac{x}{x-4} = (x-4)2+(x-4)\left(\frac{4}{x-4}\right)$$

Distribute.

$x = 2x - 8 + 4$ Solve the first-degree equation.

$-x = -4$

$x = 4$ This is the proposed solution.

Check: When $x = 4$, the denominator equals 0! Recall from your earlier work that 4 is a restricted value and that x cannot equal 4. Thus, this equation has no solution, which is written $\varnothing$.

d) $\dfrac{2}{x-2}+\dfrac{1}{x^2-4} = \dfrac{3}{x+2}$

Because $x^2 - 4 = (x - 2)(x + 2)$, the LCD $= (x - 2)(x + 2)$.

$$(x-2)(x+2)\left(\frac{2}{x-2}+\frac{1}{x^2-4}\right) = (x-2)(x+2)\left(\frac{3}{x+2}\right)$$

$$(x-2)(x+2)\left(\frac{2}{x-2}\right)+(x-2)(x+2)\left(\frac{1}{(x-2)(x+2)}\right) = (x-2)(x+2)\left(\frac{3}{x+2}\right)$$

$(x + 2)(2) + 1 = (x - 2)(3)$ Solve the first-degree equation.

$2x + 4 + 1 = 3x - 6$

$-x = -11$

$x = 11$ This is the proposed solution.

Check: Try the check on your own. The solution checks in the original equation.

e) $\dfrac{2x}{x-3} - \dfrac{x}{x+2} = \dfrac{-6}{x^2 - x - 6}$

$$(x-3)(x+2)\left(\frac{2x}{x-3} - \frac{x}{x+2}\right) = (x-3)(x+2)\left(\frac{-6}{(x-3)(x+2)}\right)$$

Because $x^2 - x - 6 = (x - 3)(x + 2)$, the LCD $= (x - 3)(x + 2)$.

$$(x-3)(x+2)\left(\frac{2x}{x-3}\right) - (x-3)(x+2)\left(\frac{x}{x+2}\right) = (x-3)(x+2)\left(\frac{-6}{(x-3)(x+2)}\right)$$

$(x + 2)(2x) - (x - 3)(x) = -6$ Simplify each side of the equation.

$2x^2 + 4x - x^2 + 3x = -6$ This is a quadratic equation.

$x^2 + 7x + 6 = 0$ Get 0 alone on one side of the equation.

$(x + 6)(x + 1) = 0$ Factor.

$x + 6 = 0$ or $x + 1 = 0$ Set each factor to 0 to solve.

$x = -6$ or $x = -1$ These are the proposed solutions.

Check: Both –6 and –1 check in the original equation and are solutions.

f) $\dfrac{x+3}{x^2 - 3x - 4} - \dfrac{2}{x^2 + 3x + 2} = \dfrac{8}{x^2 - 2x - 8}$ Factor the denominators to find the LCD.

$\dfrac{x+3}{(x-4)(x+1)} - \dfrac{2}{(x+2)(x+1)} = \dfrac{8}{(x-4)(x+2)}$ LCD $= (x - 4)(x + 1)(x + 2)$.

$$(x-4)(x+1)(x+2)\left[\frac{x+3}{(x-4)(x+1)} - \frac{2}{(x+2)(x+1)}\right]$$

$$= (x-4)(x+1)(x+2)\left[\frac{8}{(x-4)(x+2)}\right]$$

Multiply both sides of the equation by the LCD.

$$(x-4)(x+1)(x+2)\left(\frac{x+3}{(x-4)(x+1)}\right)-(x-4)(x+1)(x+2)\left(\frac{2}{(x+2)(x+1)}\right)$$

$$= (x-4)(x+1)(x+2)\left(\frac{8}{(x-4)(x+2)}\right)$$

Use the distributive property to multiply the LCD times each term.

$(x + 2)(x + 3) - (x - 4)(2) = (x + 1)8$ — Simplify both sides of the equation.

$x^2 + 5x + 6 - 2x + 8 = 8x + 8$ — Solve the quadratic equation.

$x^2 + 3x + 14 = 8x + 8$ — Combine similar terms.

$x^2 - 5x + 6 = 0$ — Get 0 alone on one side.

$(x - 2)(x - 3) = 0$ — Factor and set each factor equal to 0 to solve.

$x - 2 = 0$ or $x - 3 = 0$

$x = 2$ or $x = 3$ — These are the proposed solutions.

Check: Both 2 and 3 check when they are substituted into the original equation.

g) $\dfrac{2}{x-3}-\dfrac{3}{x+3}=\dfrac{12}{x^2-9}$

$$(x-3)(x+3)\frac{2}{x-3}-(x-3)(x+3)\frac{3}{x+3}=\frac{12}{(x-3)(x+3)}(x-3)(x+3)$$

Multiply both sides by the LCD: $(x - 3)(x + 3)$.

$2(x + 3) - 3(x - 3) = 12$ — Divide out equal factors.

$2x + 6 - 3x + 9 = 12$ — Distribute.

$-x + 15 = 12$ — Combine similar terms.

$-x = -3$ — Subtract 15 from both sides.

$x = 3$

Check: Notice that when $x = 3$, two denominators in the original equation will equal 0. Therefore, there is no solution to the equation, written $\varnothing$.

h) $\frac{r}{r-1}=\frac{2}{r^2-1}$

$(r-1)(r+1)\left(\frac{r}{r-1}\right)=(r-1)(r+1)\left(\frac{2}{(r-1)(r+1)}\right)$ Multiply both sides of the equation by the LCD = $(r-1)(r+1)$.

$(r+1)r=2$ Distribute.

$r^2+r=2$ Solve the quadratic equation.

$r^2+r-2=0$ Get 0 alone on one side of the equation.

$(r+2)(r-1)=0$ Factor.

$r+2=0$ or $r-1=0$ Set each factor equal to 0 to solve.

$r=-2$ or $r=1$ These are the proposed solutions.

Check: $r = 1$ makes the denominator of the original equation equal 0, so 1 is not a solution. Recall from our earlier work that 1 is called a **restricted value** and that r cannot equal 1.

If $r=-2$, $\frac{-2}{(-2)-1}\stackrel{?}{=}\frac{2}{(-2)^2-1}$

$\frac{-2}{-3}=\frac{2}{3}$

So –2 is a solution to the original equation.

7.6 APPLICATIONS

In this section, we solve three basic types of applications that lead to equations with rational expressions: number problems, motion problems, and work problems. In setting up the number problems, remember that the numerator is the top of the fraction, the denominator is the bottom of the fraction, and the reciprocal of a number a is $\frac{1}{a}$.

NUMBER PROBLEMS

Use the following number problems to improve your translation skills from English to Mathematics. Then, use your rational equation solving skills to solve the resulting equation. Remember to answer the question from the original problem.

Exercise 7.13

If twice the reciprocal of a number is subtracted from the number, the result is $\frac{7}{3}$. Find the number.

Solution 7.13

Let a = original number.

Then $\frac{1}{a}$ = the reciprocal.

Twice the reciprocal = $2\left(\frac{1}{a}\right)$, so to subtract the reciprocal from the number (a), we write:

$a-2\left(\frac{1}{a}\right) = \frac{7}{3}$ Write an equation.

$a-\left(\frac{2}{a}\right) = \frac{7}{3}$ LCD = $3a$.

$3a\left(a-\frac{2}{a}\right) = 3a\left(\frac{7}{3}\right)$ Multiply both sides of the equation by $3a$.

$3a^2 - 6 = 7a$ Solve the quadratic equation.

$3a^2 - 7a - 6 = 0$ Get 0 alone on one side of the equation.

$(3a + 2)(a - 3) = 0$ Factor and set each factor equal to 0.

$3a + 2 = 0$ or $a - 3 = 0$ Solve each first-degree equation.

$3a = -2$

$a = -\frac{2}{3}$ or $a = 3$

If $a = -\frac{2}{3}, \frac{1}{a} = -\frac{3}{2}$ If $a = 3, \frac{1}{a} = \frac{1}{3}$

Check:

For $a = -\frac{2}{3}$

$-\frac{2}{3}-2\left(-\frac{3}{2}\right)=\frac{7}{3}$

$-\frac{2}{3}+3=\frac{7}{3}$

$\frac{7}{3}=\frac{7}{3}$

For $a = 3$

$3-2\left(\frac{1}{3}\right)=\frac{7}{3}$

$3-\frac{2}{3}=\frac{7}{3}$

$\frac{7}{3}=\frac{7}{3}$

Substitute each value into the original equation.

Both solutions check.

MOTION PROBLEMS

Motion problems involve a moving object, often a plane, car, boat, or a person walking or riding a bicycle. These problems may ask you to account for the rate of the object when it is affected by the speed of the wind (for the plane or bicyclist) or the speed of the current (for the boat). Remember that an object moves faster *with* the wind or *with* the current (downstream). The following table contains some sample language:

Boat	Plane	Symbol
Speed in still water	Speed in still air	x
Current	Wind speed	y
Speed upstream	Speed against wind	$x - y$
Speed downstream	Speed with wind	$x + y$

These problems make use of the formula $D = R \cdot T$, where D is distance, R is the rate, and T is time.

Exercise 7.14

A boat travels 20 miles up the river in the same amount of the time it takes to travel 40 miles down the river. If the current is 10 miles per hour, what is the speed of the boat in still water?

Solution 7.14

Let x = speed of the boat in still water.

	R	*T*	*D*
Upstream	$x - 10$		20
Downstream	$x + 10$		40

$R \cdot T = D$ Solve for T.

$\frac{R \cdot T}{R} = \frac{D}{R}$ Divide both sides by R.

$T = \frac{D}{R}$

Now, we can complete the table:

	R	*T*	*D*
Upstream	$x - 10$	$\frac{20}{x-10}$	20
Downstream	$x + 10$	$\frac{40}{x+10}$	40

Because the time upstream must be equal to the time downstream:

$$\frac{20}{x-10} = \frac{40}{x+10}$$

$$(x-10)(x+10)\left(\frac{20}{x-10}\right)$$

$$= (x-10)(x+10)\left(\frac{40}{x+10}\right)$$ Multiply by LCD.

$(x + 10)(20) = (x - 10)(40)$ Solve the first-degree equation.

$20x + 200 = 40x - 400$ Subtract 200 and $40x$ from both sides.

$-20x = -600$

$$\frac{-20x}{-20} = \frac{-600}{-20}$$ Divide both the sides of the equation by –20.

$x = 30$ miles per hour This is the speed of the boat in still water.

Check:

$$\frac{20}{x-10} = \frac{40}{x+10}$$ This is the original equation.

$$\frac{20}{(30)-10} = \frac{40}{(30)+10}$$ Substitute 30 into the original equation.

$$\frac{20}{20} = \frac{40}{40}$$ 30 checks.

WORK PROBLEMS

Work problems involve the length of the time to do a job and usually involve two or more people or machines working at a task such as painting or typing. These techniques are also used for pipes that fill and drain sinks or swimming pools. Study the following table that relates total job time to the fractional part of the job completed in 1 hour:

Time to Complete Job	Part of Job Completed in 1 Hour	Part of Job Completed in 2 Hours
5 hours	$\frac{1}{5}$	$\frac{4}{5}$
8 hours	$\frac{1}{5}$	$\frac{4}{5}$
x hours	$\frac{4}{5}$	$\frac{4}{5}$

Exercise 7.15

Joan can type a manuscript in 30 hours. Lisa can do the job in 60 hours. How long would it take them if they work together?

Solution 7.15

Let x = time to complete the job working together.

In 1 hour:

* Joan completes $\frac{1}{30}$ of the job.
* Lisa completes $\frac{1}{60}$ of the job.
* Together, they complete $\frac{1}{x}$ of the job.

So, $\frac{1}{30} + \frac{1}{60} = \frac{1}{x}$

$$60x\left(\frac{1}{30}+\frac{1}{60}\right) = 60x\left(\frac{1}{x}\right) \qquad \text{Multiply by the LCD.}$$

$$60x\left(\frac{1}{30}\right)+60x\left(\frac{1}{60}\right) = 60x\left(\frac{1}{x}\right) \qquad \text{Distribute.}$$

$2x + x = 60$ Solve.

$3x = 60$

$x = 20$ hours

Exercise 7.16

A cold-water faucet can fill a sink in 15 minutes, and a hot-water faucet can fill it in 10 minutes. The drain can empty the sink in 18 minutes. If both faucets are on and the drain is accidentally left open, how long will it take to fill the sink?

Solution 7.16

Let x = time to fill the sink.

In 1 minute:

The cold-water faucet fills $\frac{1}{15}$ of the sink.

The hot-water faucet fills $\frac{1}{10}$ of the sink.

The drain empties $\frac{1}{18}$ of the sink.

Then $\frac{1}{15}+\frac{1}{10}-\frac{1}{18}=\frac{1}{x}$

$90x\left(\frac{1}{15}+\frac{1}{10}-\frac{1}{18}\right)=90x\left(\frac{1}{x}\right)$ Multiply by the LCD.

$90x\left(\frac{1}{15}\right)+90x\left(\frac{1}{10}\right)-90x\left(\frac{1}{18}\right)=90x\left(\frac{1}{x}\right)$ Distribute.

$6x + 9x - 5x = 90$ Solve.

$10x = 90$

$x = 9$ minutes

It takes 9 minutes to fill the sink.

7.7 COMPLEX FRACTIONS

A **complex fraction** is a rational expression that contains fractions in the numerator and/or denominator. There are two methods used to simplify complex functions. If your instructor uses only Method 1, study Exercise 7.17. If your instructor uses only Method 2, study Exercise 7.18. If you want to compare these two methods, study Exercise 7.19.

Method 1:

1. Write the numerator of the complex fraction as a single fraction.
2. Write the denominator of the complex fraction as a single fraction.
3. Perform the indicated division by inverting the divisor and multiplying.
4. Reduce, if possible.

Method 2 (sometimes called the LCD method):

1. Find the least common denominator (LCD) of *all* the denominators within the complex fraction.
2. Multiply the numerator and denominator of the complex fraction by the LCD.
3. Reduce, if possible.

SIMPLIFYING COMPLEX FRACTIONS USING METHOD 1

This method is sometimes called the invert and multiply method. Use your adding and subtracting rational expression skills to simplify the numerator and then the denominator before attempting to invert and multiply.

Exercise 7.17

Simplify each complex fraction:

a) $$\frac{\frac{2}{5}}{4}$$

b) $$\frac{\frac{3x^2}{y}}{\frac{9x^6}{10y^3}}$$

c) $$\frac{\frac{m+1}{m^2-9}}{\frac{4m+4}{m^2+5m+6}}$$

Solution 7.17

a) $$\frac{\frac{2}{5}}{4}$$

$$= \frac{\frac{2}{5}}{\frac{4}{1}}$$

The numerator and denominator are single fractions.

$$= \frac{2}{5} \cdot \frac{1}{4}$$

Invert and multiply.

$$= \frac{\mathbf{2} \cdot 1}{5 \cdot \mathbf{2} \cdot 2}$$

Reduce by dividing out common factors.

$$= \frac{1}{10}$$

b) $$\frac{\frac{3x^2}{y}}{\frac{9x^6}{10y^3}}$$

The numerator and denominator are single fractions.

$$= \frac{3x^2}{y} \cdot \frac{10y^3}{9x^6}$$

Invert and multiply.

$$= \frac{3x^2 \cdot 10y^3}{y \cdot 3 \cdot 3x^6}$$

Reduce by dividing out the common factors: 3, x^2 and y.

$$= \frac{10y^2}{3x^4}$$

c) $$\frac{\frac{m+1}{m^2-9}}{\frac{4m+4}{m^2+5m+6}}$$

Numerator and denominator are single fractions.

$$= \frac{m+1}{m^2-9} \cdot \frac{m^2+5m+6}{4m+4}$$

Invert and multiply.

$$= \frac{m+1}{(m+3)(m-3)} \cdot \frac{(m+2)(m+3)}{4(m+1)}$$

Divide out common factors: $m + 1$ and $m + 3$.

$$= \frac{m+2}{4(m-3)}$$

SIMPLIFYING COMPLEX FRACTIONS USING METHOD 2

This method is sometimes called the LCD method. You may want to write any whole number as a fraction by writing the whole number divided by 1 to help you identify the terms within the complex fraction (see Example (a)).

Exercise 7.18

Simplify each complex fraction:

a) $\dfrac{\frac{2}{a}-3}{\frac{2}{a}+3}$

b) $\dfrac{\frac{4}{p^2q}-\frac{3}{pq^2}}{\frac{5}{p^2q^2}+\frac{1}{pq}}$

Solution 7.18

a) $\dfrac{\frac{2}{a}-3}{\frac{2}{a}+3}=\dfrac{a\left(\frac{2}{a}-\frac{3}{1}\right)}{a\left(\frac{2}{a}+\frac{3}{1}\right)}$ Multiply the numerator and denominator by the LCD: a.

$=\dfrac{a\left(\frac{2}{a}\right)-a\left(\frac{3}{1}\right)}{a\left(\frac{2}{a}\right)+a\left(\frac{3}{1}\right)}$ Use the distributive property to multiply the LCD times each term.

$=\dfrac{2-3a}{2+3a}$ $a\left(\frac{2}{a}\right)=\frac{2a}{a}=2$.

b) $\dfrac{\frac{4}{p^2q}-\frac{3}{pq^2}}{\frac{5}{p^2q^2}+\frac{1}{pq}}=\dfrac{p^2q^2\left(\frac{4}{p^2q}-\frac{3}{pq^2}\right)}{p^2q^2\left(\frac{5}{p^2q^2}+\frac{1}{pq}\right)}$ Multiply the numerator and denominator by the LCD: p^2q^2.

$$=\frac{p^2q^2\left(\frac{4}{p^2q}\right)-p^2q^2\left(\frac{3}{pq^2}\right)}{p^2q^2\left(\frac{5}{p^2q^2}\right)+p^2q^2\left(\frac{1}{pq}\right)}$$

Use the distributive property to multiply the LCD times each term.

$$=\frac{4q-3p}{5+pq}$$

SIMPLIFYING COMPLEX FRACTIONS: A COMPARISON OF METHODS 1 AND 2

Generally, either method can be used to simplify a complex fraction. The following exercise demonstrates both methods. Which method do you prefer?

Exercise 7.19

Simplify:

$$\frac{2+\frac{3}{4}}{3-\frac{1}{8}}$$

Solution 7.19

Method 1:

$$\frac{2+\frac{3}{4}}{3-\frac{1}{8}}=\frac{\frac{2}{1}\left(\frac{4}{4}\right)+\frac{3}{4}}{\frac{3}{1}\left(\frac{8}{8}\right)-\frac{1}{8}}$$

Write the numerator as a single fraction. Write the denominator as a single fraction.

$$=\frac{\frac{8}{4}+\frac{3}{4}}{\frac{24}{8}-\frac{1}{8}}$$

Find the LCD for the numerator (4) and rewrite each fraction with the LCD. Find the LCD for the denominator (8) and rewrite each fraction with the LCD.

$$=\frac{\frac{11}{4}}{\frac{23}{8}}$$

Add the fractions in the numerator. Subtract the fractions in the denominator.

$$=\frac{11}{4}\cdot\frac{8}{23}$$

Invert and multiply.

$$=\frac{22}{23}$$

Method 2:

$$\frac{2+\frac{3}{4}}{3-\frac{1}{8}}=\frac{8\left(\frac{2}{1}+\frac{3}{4}\right)}{8\left(\frac{3}{1}-\frac{1}{8}\right)}$$

Find the LCD of all the denominators: The LCD of 1, 4, and 8 is 8. Use the distributive property to multiply the LCD times each term.

$$8\left(\frac{2}{1}\right)=\frac{16}{1}=16$$

$$8\left(\frac{3}{4}\right)=\frac{24}{4}=6$$

$$8\left(\frac{3}{1}\right)=\frac{24}{1}=24$$

$$8\left(\frac{1}{8}\right)=\frac{8}{8}=1$$

$$\frac{16+6}{24-1}=\frac{22}{23}$$

7.8 RATIOS AND PROPORTIONS

Many word problems can be set up using proportions. If you studied geometry, recall that one of the properties of similar triangles is that corresponding sides are proportional.

RATIOS

A **ratio** provides a way of comparing two quantities with the same units. The ratio of two numbers a and b is written as $\frac{a}{b}$ or $a : b$ or a to b. Reducing a ratio is equivalent to reducing a fraction.

Exercise 7.20

Write each word phrase as a ratio. Write each ratio in lowest terms:

a) 8 hours to 4 hours

b) 20 miles : $\frac{3}{4}$ mile

c) 6 days : 12 hours

Solution 7.20

a) 8 hours to 4 hours $= \dfrac{8\,\text{hours}}{4\,\text{hours}}$ — Write the ratio as a fraction.

$= \dfrac{2\,\text{hours}}{1\,\text{hour}}$ — Reduce the fraction.

We would ordinarily write $\dfrac{2}{1}$ as 2, but with ratios, we generally write denominators of 1.

b) 20 miles: $\dfrac{3}{4}$ mile $= \dfrac{20\,\text{miles}}{\frac{3}{4}\,\text{mile}}$ — Write the ratio as a fraction.

$= \dfrac{\frac{20}{1}}{\frac{3}{4}}$ — Simplify the complex fraction. It is easier to work without the units (miles); just remember to put them in the answer.

$= \dfrac{20}{1} \cdot \dfrac{4}{3}$ — Invert and multiply.

$= \dfrac{80}{3}$ — Multiply the numerators; multiply the denominators.

$= \dfrac{80\,\text{miles}}{3\,\text{miles}}$ — Put the units back in the numerator and denominator.

c) 6 days : 12 hours = 6 days : $\dfrac{1}{2}$ day — The units must be the same. Rewrite 12 hours as $\dfrac{1}{2}$ day.

$= \dfrac{6}{\frac{1}{2}}$ — Write the ratio as a complex fraction.

$= \dfrac{\frac{6}{1}}{\frac{1}{2}}$ — Simplify the complex fraction.

$= \dfrac{6}{1} \cdot \dfrac{2}{1}$ — Invert and multiply.

$= \dfrac{12}{1}$ — Multiply the numerators, multiply the denominators.

$= \dfrac{12\,\text{days}}{1\,\text{day}}$ Put the units back in the numerator and denominator.

PROPORTIONS

An equation that consists of two equal ratios is called a **proportion**. If $\frac{a}{b}$ and $\frac{c}{d}$ are two equal ratios, then $\frac{a}{b} = \frac{c}{d}$ is called a proportion. In the proportion $\frac{a}{b} = \frac{c}{d}$, a, b, c, and d are called **terms** of the proportion. Also, b and c are called the **means** of the proportion, and a and d are called the **extremes** of the proportion. Use the following formula when working with proportions:

If $\frac{a}{b} = \frac{c}{d}$, then $ad = bc$.

In words, the product of the extremes equals the product of the means.

Exercise 7.21

Solve and check each proportion:

a) $\dfrac{x}{6} = \dfrac{8}{4}$

b) $\dfrac{x}{2} = \dfrac{3}{4}$

c) $\dfrac{8}{r+24} = \dfrac{2}{r}$

Solution 7.21

a) $\dfrac{x}{6} = \dfrac{8}{4}$ The extremes are x and 4. The means are 6 and 8.

$4x = 6 \cdot 8$ The product of the extremes equals the product of the means.

$4x = 48$ Solve the equation.

$\dfrac{4x}{4} = \dfrac{48}{4}$ Divide both sides by 4.

$x = 12$

Check:

$\frac{x}{6}=\frac{8}{4}$ This is the original proportion.

$\frac{12}{6}\overset{?}{=}\frac{8}{4}$ Substitute 12 into the original proportion.

$2=2$ 12 checks.

b) $\frac{x}{2}=\frac{3}{4}$ The extremes are x and 4. The means are 2 and 3.

$4x=2\cdot 3$ The product of the extremes equals the product of the means.

$4x=6$ Solve the equation.

$x=\frac{3}{2}$ Divide both the sides by 4.

$x=\frac{3}{2}$ Reduce the fraction.

Check:

$\frac{x}{2}=\frac{3}{4}$ This is the original proportion.

$\frac{\frac{3}{2}}{2}\overset{?}{=}\frac{3}{4}$ Substitute $\frac{3}{2}$ into the original proportion.

$\frac{\frac{3}{2}}{\frac{2}{1}}=\frac{3}{4}$ Simplify the complex fraction: $\frac{3}{2}\cdot\frac{1}{2}=\frac{3}{4}$.

$\frac{3}{4}=\frac{3}{4}$ $\frac{3}{2}$ checks.

c) $\frac{8}{r+24}=\frac{2}{r}$ The extremes are 8 and r. The means are $(r+24)$ and 2.

$8r = 2(r + 24)$	The product of the extremes equals the product of the means.
$8r = 2r + 48$	Solve the equation.
$6r = 48$	Subtract $2r$ from both sides of the equation.
$\frac{6r}{6} = \frac{48}{6}$	Divide both sides of the equation by 6.
$r = 8$	
Check:	
$\frac{8}{r+24} = \frac{2}{r}$	This is the original proportion.
$\frac{8}{(8)+24} \stackrel{?}{=} \frac{2}{8}$	Substitute 8 into the original proportion.
$\frac{8}{32} = \frac{2}{8}$	Simplify the fraction.
$\frac{1}{4} = \frac{1}{4}$	8 checks.

7.9 VARIATION

Two quantities vary directly when one quantity is a constant multiple of the other. For example, if you buy 3 cans of cat food for 2 cats, you would buy 6 cans of cat food for 4 cats. (Two times as many cats means two times as many cans of cat food.) The constant multiple is usually represented by k. In an indirect variation, one quantity increases as the other decreases (for example, as the number of credits you take increases, your free time decreases). Some types of variation are as follows:

Type	Language	Symbols
Direct	y varies directly as x	$y = kx$
Indirect	y varies inversely as x or y varies indirectly as x	$y = \frac{k}{x}$

To solve variation problems, do the following:

1. Decide which basic formula applies (direct $y = kx$; indirect $y = \frac{k}{x}$).
2. Assign variables, if not already stated in the problem.
3. Solve for k, the constant of proportionality.
4. Use the value of k in the basic formula to solve for the requested unknown.

Exercise 7.22

If t varies directly as s, and $t = 9$ when $s = 6$, find t when s is $\frac{2}{5}$.

Solution 7.22

Find k first:

$t = ks$ — Use direct variation.

$9 = k(6)$ — Substitute $t = 9$, $s = 6$.

$\frac{9}{6} = k$ — Solve for k.

$\frac{3}{2} = k$

Find t:

$t = \frac{3}{2}s$ — Use direct variation $k = \frac{3}{2}$.

$t = \frac{3}{2}\left(\frac{2}{5}\right)$ — Substitute $s = \frac{2}{5}$.

$t = \frac{3}{5}$ — Solve for t.

Exercise 7.23

If p varies indirectly as q^2, and $p = 12$ when $q = 2$, find p when $q = 4$.

Solution 7.23

First find k:

$p = \frac{k}{q^2}$ — Use indirect variation.

$12 = \frac{k}{(2)^2}$ — Substitute $p = 12$, $q = 2$.

$12 = \dfrac{k}{4}$	Solve for k.
$4(12) = 4\left(\dfrac{k}{4}\right)$	Multiply by the LCD.
$48 = k$	Now use k to find p when $q = 4$.
$p = \dfrac{48}{q^2}$	Use indirect variation: $k = 48$.
$p = \dfrac{48}{(4)^2}$	Substitute $q = 4$.
$p = \dfrac{48}{16}$	
$p = 3$	This is the solution.

Exercise 7.24

The force F required to stretch a spring varies directly as the distance d that the spring is stretched. If a force of 5 pounds stretches a spring 3 inches, find:

a) k

b) The force needed to stretch the spring 6 inches. (Note that this answer will be in pounds.)

Solution 7.24

a) $F = kd$	Use the variables given in the problem in the formula for direct variation.
$5 = k(3)$	Substitute $F = 5$, $d = 3$ into the formula.
$\dfrac{5}{3} = k$	Divide both sides by 3 to solve for k.
b) $F = \dfrac{5}{3}d$	Use the value of k found from part (a).
$F = \dfrac{5}{3}(6)$	Substitute 6 for d.
$F = 10$ pounds	

TEST YOURSELF

1) State the restrictions on the variable in the following rational expressions:

(a) $\dfrac{2x-3}{8}$

(b) $\dfrac{x+3}{x-7}$

(c) $\dfrac{9x^2}{5x+6}$

(d) $\dfrac{11x}{x^2-3x+2}$

2) Reduce to the lowest terms:

(a) $\dfrac{3x+6}{7x+14}$

(b) $\dfrac{x^2+3x+2}{x^2-x-2}$

(c) $\dfrac{4x^2+4x+1}{6x^2+7x+2}$

(d) $\dfrac{x^2+xy+xz+yz}{2x^2+3xz+2xy+3yz}$

3) Write in the lowest terms:

(a) $\dfrac{5a-b}{b-5a}$

(b) $\dfrac{a-2b}{2b+a}$

(c) $\dfrac{x^2-36}{6-x}$

(d) $\dfrac{x^2+x-2}{-x^2+1}$

4) Perform the indicated operations:

(a) $\frac{3x^2}{27x^3} \cdot \frac{12x}{16x}$

(b) $\frac{x^2-4}{2y} \div \frac{2-x}{6xy}$

(c) $\frac{a^2+5a+6}{2a^2+9a+4} \cdot \frac{2a^2+7a-4}{2a^2+7a+6}$

(d) $\frac{a^2+3a+2}{a^2+5a+6} \div \frac{a^2+2a+1}{a^2+4a+3}$

5) Find the least common multiple (LCM):

(a) $6x^3y$ and $8xy^2$

(b) $x^2 + 4x + 3$ and $x^2 + 5x + 6$

6) Find the least common denominator (LCD):

(a) $\frac{2}{x^2-x-6}$ and $\frac{4}{x^2-5x+6}$

(b) $\frac{1}{8}$ and $\frac{1}{14}$

7) Perform the indicated operations:

(a) $\frac{3}{y^2} + \frac{8}{y^2}$

(b) $\frac{4x+3}{2x+7} + \frac{3x-8}{2x+7}$

(c) $\frac{2}{m^2-4} - \frac{3}{m+2}$

(d) $\frac{6z}{(z-1)^2} - \frac{2z}{z^2-1}$

8) Solve:

(a) $\dfrac{x}{x+4}=\dfrac{2}{x+4}+5$

(b) $\dfrac{2}{x-3}-\dfrac{3}{x+3}=\dfrac{12}{x^2-9}$

(c) $\dfrac{1}{2x^2-x-1}+\dfrac{2}{2x^2+3x+1}=\dfrac{3}{x^2-1}$

(d) $\dfrac{2}{x}+\dfrac{20}{x^2+5x}=\dfrac{x+1}{x+5}$

(e) $\dfrac{x+3}{x^2-x}=\dfrac{8}{x^2-1}$

9) Solve:

(a) The sum of a number and twice its reciprocal is $\dfrac{27}{5}$. Find the number.

(b) Bill can type 5 forms of a test in 4 hours. Lisa can type the same 5 forms in 6 hours. If they work together, how long would it take to type all 5 forms?

(c) Triangles *ABC* and *XYZ* are similar. The lengths of the corresponding sides are a and x, b and y, and c and z. If $a = 10$, $x = 15$, $y = 12$, and $z = 14$, find b.

(d) One number is twice another. The sum of their reciprocals is $\dfrac{15}{2}$. Find the two numbers.

(e) The sum of a number and its reciprocal is $\dfrac{10}{4}$. Find the number.

(f) The sum of a number and three times its reciprocal is $\dfrac{13}{2}$. Find the number.

(g) With a wind of 20 miles per hour, an airplane can travel 1,400 miles. In the same time, the plane can travel 1,200 miles against the same wind. Find the speed of the plane in still air.

(h) With a wind of 15 miles per hour, an airplane can travel 1,325 miles. In the same time, the plane can travel 1,175 miles against the same wind. Find the speed of the plane in still air.

(i) Christian can type five forms of a test in 4 hours. Pekins can type the same five forms in 6 hours. If they work together, how long would it take to type all five forms?

(j) A water tank can be filled in 6 hours by an inlet pipe and emptied in 9 hours by an outlet pipe. How long will it take to fill the tank if both pipes are left open?

(k) A water tank can be filled in 12 hours by an inlet pipe and emptied in 20 hours by an outlet pipe. How long will it take to fill the tank if both pipes are left open?

10) Simplify:

(a) $\dfrac{\dfrac{36x^4}{5y^4z^5}}{\dfrac{9xy^2}{15z^5}}$

(b) $\dfrac{\dfrac{9}{x}+\dfrac{3}{x^2}}{\dfrac{4}{5}+\dfrac{1}{x}}$

(c) $\dfrac{\dfrac{4x+8}{3x^2}}{\dfrac{4}{6x}}$

11) Solve each proportion:

(a) $\dfrac{x}{7}=\dfrac{3}{21}$

(b) $\dfrac{3}{5}=\dfrac{6}{x}$

(c) $\dfrac{a}{4}=\dfrac{7}{5}$

(d) $\dfrac{6}{x+18}=\dfrac{2}{x}$

(e) $\dfrac{5}{x^2-25}=\dfrac{1}{x-5}$

12) For each problem, assume y varies directly as x:

(a) If $y = 12$ when $x = 4$, find y when $x = 20$.

(b) If $y = 15$ when $x = 10$, find y when $x = 8$.

13) For each problem, assume y varies directly as the square root of x:

(a) If $y = 5$ when $x = 25$, find y when $x = 16$.

(b) If $y = 10$ when $x = 16$, find y when $x = 36$.

14) For each problem, assume y varies indirectly as x:

    (a) If $y = 2$ when $x = 8$, find y when $x = 20$.

    (b) If $y = 8$ when $x = 3$, find y when $x = 12$.

15) Solve each variation problem:

    (a) The force F of a spring varies directly as the distance d the spring is compressed. If it takes a force of 10 pounds to compress a spring 1 foot, find the force necessary to compress the spring 2 feet.

    (b) The weight F of a body varies inversely as the square of its distance d from the center of the Earth. If a space module weighs 10 tons when it is 4,000 miles from the center of the Earth, how much does it weigh when it is 4,800 miles from the center of the Earth?

TEST YOURSELF ANSWERS

1) (a) None

(b) $x \neq 7$

(c) $x \neq -\frac{6}{5}$

(d) $x \neq 1, 2$

2) (a) $\frac{3}{7}$

(b) $\frac{x+2}{x-2}$

(c) $\frac{2x+1}{3x+2}$

(d) $\frac{x+z}{2x+3z}$

3) (a) -1

(b) $\frac{a-2b}{2b+a}$

(c) $-x - 6$

(d) $-\frac{(x+2)}{(x+1)}$

4) (a) $\frac{1}{12x}$

(b) $-3x^2 - 6$

(c) $\frac{(a+3)(2a-1)}{(2a+1)(2a+3)}$

(d) 1

5) (a) $24x^3y^2$

(b) $(x + 1)(x + 2)(x + 3)$

6) (a) $(x + 2)(x - 2)(x - 3)$

(b) 56

7) (a) $\frac{11}{y^2}$

(b) $\frac{7x-5}{2x+7}$

(c) $\frac{8-3m}{m^2-4}$

(d) $\frac{4z(z+2)}{(z-1)^2(z+1)}$

8) (a) $-\frac{11}{2}$

(b) $\varnothing$ (3 is a restricted value)

(c) $-\frac{4}{3}$

(d) 6 (-5 is a restricted value)

(e) 3 (1 is a restricted value)

9) (a) $5, \frac{1}{5}$

(b) 2.4 hours

(c) 8

(d) $\frac{1}{5}, \frac{2}{5}$

(e) $2, \frac{1}{2}$

(f) $6, \frac{1}{2}$

(g) 260 mph

(h) 250 mph

(i) $2\frac{2}{5}$ hours

(j) 18 hours

(k) 30 hours

10) (a) $\frac{12x^3}{y^6}$

(b) $\frac{15(3x+1)}{x(4x+5)}$

(c) $\frac{2x+4}{x}$

11) (a) 1

(b) 10

(c) $\frac{28}{5}$

(d) 9

(e) 0

12) (a) 60

(b) 12

13) (a) 4

(b) 15

14) (a) $\frac{4}{5}$

(b) 2

15) (a) 20 pounds

(b) Approximately 6.9 tons

CHAPTER 8

Radical Expressions

This chapter includes simplifying radicals, performing the basic operations on radicals including addition, subtraction, multiplication, and division, and ends with solving equations containing radical expressions.

8.1 INTRODUCTION TO ROOTS AND RADICALS

We use exponents as shorthand for multiplication when we write $7^2 = 49$. We use a radical sign to reverse this process. You may want to review the exponent work covered in Chapter 5 before proceeding.

THE BASICS OF ROOTS AND RADICALS

In the expression $\sqrt{49}$, read "the square root of 49," 49 is called the **radicand** and $\sqrt{\ }$ is called a **radical sign**. Because $7^2 = 49$ and $(-7)^2 = 49$, we use:

$\sqrt{49}$ to indicate that we want the positive square root of $49 = 7$

and

$-\sqrt{49}$ to indicate that we want the negative square root of $49 = -7$.

Exercise 8.1

Find each square root:

a) $\sqrt{36}$

b) $-\sqrt{36}$

c) $\sqrt{\dfrac{25}{49}}$

d) $\dfrac{-\sqrt{36}}{-\sqrt{64}}$

Solution 8.1

a) $\sqrt{36} = 6$

No sign in front of the radical means we want the positive square root of 36: $6^2 = 36$.

b) $-\sqrt{36} = -6$

The negative sign in front of the radical means we want the negative square root of 36, $(-6)^2$.

c) $\sqrt{\dfrac{25}{49}} = \dfrac{5}{7}$

No sign in front of the radical means we want the positive square root of $\dfrac{25}{49}$: $\left(\dfrac{5}{7}\right)^2 = \dfrac{25}{49}$

d) $\dfrac{-\sqrt{36}}{-\sqrt{64}} = \dfrac{-6}{-8}$

$-\sqrt{36} = -6$; $-\sqrt{64} = -8$.

$= \dfrac{3}{4}$

Reduce the fraction.

RATIONAL VERSUS IRRATIONAL VERSUS NOT REAL

The square root of a perfect square (that is, a number like 16 that can be written as 4^2) is a rational number. The square root of any other positive number is an irrational number. The square root of *any* negative number is *not* a real number (because there are no real numbers that multiply times themselves to equal a negative number). Study the following table:

Number	Rational or Irrational	Real or Not Real
$\sqrt{36}$	Rational	Real
$-\sqrt{36}$	Rational	Real
$\sqrt{11}$	Irrational	Real
$-\sqrt{11}$	Irrational	Real
$\sqrt{-4}$	Neither	Not real
$-\sqrt{36}$	Neither	Not real

OTHER ROOTS

To write the reverse of $4^3 = 64$ and $5^4 = 625$ we write $\sqrt[3]{64} = 4$, which is read "the cube root of 64 equals 4." Likewise, $\sqrt[4]{625} = 5$ is read " the fourth root of 625 equals 5."

In the expression $\sqrt[3]{64}$ 3 is called the **index,** 64 is called the **radicand,** and $\sqrt{\ }$ is called a **radical sign.** When no index is written, it is understood to be a square root. The following table will help you recognize some square roots, cube roots, and fourth roots:

Square Roots	Cube Roots	Fourth Roots
$\sqrt{1} = 1$ $\sqrt{49} = 7$	$\sqrt[3]{1} = 1$	$\sqrt[4]{1} = 1$
$\sqrt{4} = 2$ $\sqrt{64} = 8$	$\sqrt[3]{8} = 2$	$\sqrt[4]{16} = 2$
$\sqrt{9} = 3$ $\sqrt{81} = 9$	$\sqrt[3]{27} = 3$	$\sqrt[4]{81} = 3$
$\sqrt{16} = 4$ $\sqrt{100} = 10$	$\sqrt[3]{64} = 4$	$\sqrt[4]{256} = 4$
$\sqrt{25} = 5$ $\sqrt{121} = 11$	$\sqrt[3]{125} = 5$	$\sqrt[4]{625} = 5$
$\sqrt{36} = 6$ $\sqrt{144} = 12$	$\sqrt[3]{216} = 6$	$\sqrt[4]{1,296} = 6$

Exercise 8.2

Find each real root if it exists:

a) $\sqrt[3]{64}$

b) $\sqrt[3]{-8}$

c) $-\sqrt[3]{1}$

d) $\sqrt[4]{\dfrac{16}{81}}$

e) $\sqrt[4]{-256}$

Solution 8.2

a) $\sqrt[3]{64} = 4$ Because $4^3 = 64$

b) $\sqrt[3]{-8} = -2$ Because $(-2)^3 = -8$

c) $-\sqrt[3]{1} = -1$ Because $-\sqrt[3]{1} = -1(\sqrt[3]{1}) = -1(1) = -1$

d) $\sqrt[4]{\dfrac{16}{81}} = \dfrac{2}{3}$ Because $\left(\dfrac{2}{3}\right)^4 = \dfrac{16}{81}$

e) $\sqrt[4]{-256}$ This does not exist. No real number multiplied times itself four times equals a negative number.

An **even root** (square root, fourth root, sixth root, and so on) of a negative number is not a real number. An **odd root** (cube root, fifth root, seventh root, and so on) of a negative number is a real number.

8.2 SIMPLIFYING RADICAL EXPRESSIONS

To simplify radicals, we use the following rules: For x, y nonnegative real numbers,

* $\sqrt[n]{xy} = \sqrt[n]{x}\sqrt[n]{y}$ The nth root of a product is the product of the nth roots.
* $\sqrt[n]{\dfrac{x}{y}} = \dfrac{\sqrt[n]{x}}{\sqrt[n]{y}}$ The nth root of a quotient is the quotient of the nth roots.
* $\sqrt[n]{x^n} = x$ The nth root of x^n is x.

A radical is in simplified form if

* The index is greater than any exponent in the radicand.
* There are no fractions under the radical.
* There are no radicals in the denominator of a fraction.

To simplify $\sqrt[n]{*}$

1. Write the radicand as a product of the perfect n^{th} power, when possible.
2. Use $\sqrt[n]{x^n} = x$ to bring perfect factors outside the radical. Any factors that cannot be written as perfect n^{th} powers stay in the radicand.

Note: We will assume all variables represent nonnegative real numbers!

Exercise 8.3

Simplify:

a) $\sqrt{48x^4}$

b) $\sqrt[3]{48x^4}$

c) $\sqrt[4]{48x^4}$

d) $\sqrt[5]{48x^4}$

Solution 8.3

a) $\sqrt{48x^4} = \sqrt{16x^4 \cdot 3}$ — Factor 48 into 16 · 3, because 16 is a perfect square.

$= \sqrt{16x^4} \cdot \sqrt{3}$ — Use $\sqrt{x \cdot y} = \sqrt{x}\sqrt{y}$.

$= 4x^2\sqrt{3}$ — Use $\sqrt[n]{x^n} = x$.

b) $\sqrt[3]{48x^4} = \sqrt[3]{8x^3 \cdot 6x}$ — Factor 48 into 8 · 6, because 8 is a perfect cube.

$= \sqrt[3]{8x^3}\sqrt[3]{6x}$ — $\sqrt[3]{xy} = \sqrt[3]{x}\,\sqrt[3]{y}$.

$= 2x\sqrt[3]{6x}$ — Use $\sqrt[n]{x^n} = x$.

c) $\sqrt[4]{48x^4} = \sqrt[4]{16x^4 \cdot 3}$ — Factor 48 into 16 · 3, because 16 is a perfect square.

$= \sqrt[4]{16x^4} \cdot \sqrt[4]{3}$ — $\sqrt[4]{x \cdot y} = \sqrt[4]{x}\,\sqrt[4]{y}$

$= 2x\sqrt[4]{3}$ — Use $\sqrt[n]{x^n} = x$.

d) $\sqrt[5]{48x^4}$ — The index is greater than the exponents in the radicand.

$= \sqrt[5]{48x^4}$ — This cannot be simplified further.

Exercise 8.4

Simplify:

a) $\sqrt{75x^2y}$

b) $3a\sqrt{32a^3}$

c) $xy\sqrt[3]{8x^4y^3}$

Solution 8.4

a) $\sqrt{75x^2y} = \sqrt{25x^2 \cdot 3y}$ — Factor the radicand: $75 = 25 \cdot 3$.

$= \sqrt{25x^2}\sqrt{3y}$ — Use $\sqrt{xy} = \sqrt{x}\sqrt{y}$.

$= 5x\sqrt{3y}$ — Use $\sqrt[n]{x^n} = x$.

b) $3a\sqrt{32a^3} = 3a\sqrt{16a^2 \cdot 2a}$ — Factor the radicand.

$= 3a\sqrt{16a^2}\sqrt{2a}$ — $\sqrt{xy} = \sqrt{x}\sqrt{y}$.

$= 3a \cdot 4a\sqrt{2a}$ — Use $\sqrt[n]{x^n} = x$.

$= 12a^2\sqrt{2a}$ — Multiply. Be careful to write the exponent, 2, on a, and not as the index.

c) $xy\sqrt[3]{8x^4y^3} = xy\sqrt[3]{8x^3y^3x^1}$ — Factor the radicand and write each factor as a perfect cube, when possible.

$= xy\sqrt[3]{8x^3y^3}\sqrt[3]{x}$ — Use $\sqrt{xy} = \sqrt{x}\sqrt{y}$.

$= xy \cdot 2xy\sqrt[3]{x}$ — Use $\sqrt[n]{x^n} = x$.

$= 2x^2y^2\sqrt[3]{x}$ — Multiply. Be sure to write the index of 3 in your answer.

If the radicand is a fraction and can be factored as a perfect n^{th} root, we can use $\sqrt[n]{\frac{x}{y}} = \frac{\sqrt[n]{x}}{\sqrt[n]{y}}$ and $\sqrt[n]{x^n} = x$ to simplify.

Exercise 8.5

Simplify:

a) $\sqrt{\frac{4}{9}}$

b) $\sqrt[3]{\frac{8x^3}{125y^6}}$

c) $\sqrt{\frac{8}{18}}$

Solution 8.5

a) $\sqrt{\frac{4}{9}}=\frac{\sqrt{4}}{\sqrt{9}}$ Use $\sqrt{\frac{x}{y}}=\frac{\sqrt{x}}{\sqrt{y}}$.

$=\frac{2}{3}$ Use $\sqrt[n]{x^n}=x$.

b) $\sqrt[3]{\frac{8x^3}{125y^6}}=\frac{\sqrt[3]{8x^3}}{\sqrt[3]{125y^6}}$ Use $\sqrt[3]{\frac{x}{y}}=\frac{\sqrt[3]{x}}{\sqrt[3]{y}}$.

$=\frac{2x}{5y^2}$ Use $\sqrt[n]{x^n}=x$.

c) $\sqrt{\frac{8}{18}}=\sqrt{\frac{2\cdot 4}{2\cdot 9}}$ Reduce the fraction first.

$=\sqrt{\frac{4}{9}}=\frac{2}{3}$ See Part (a).

Exercise 8.6

Simplify:

a) $\sqrt{\frac{50x^3}{9}}$

b) $\sqrt{\frac{48}{16}}$

c) $\frac{4\sqrt{54a^3}}{\sqrt{25}}$

d) $\sqrt[3]{\frac{64x^5}{27}}$

e) $\sqrt{\frac{75p^4}{16m^2}}$

Solution 8.6

a) $\sqrt{\dfrac{50x^3}{9}} = \dfrac{\sqrt{50x^3}}{\sqrt{9}}$ Use $\sqrt{\dfrac{x}{y}} = \dfrac{\sqrt{x}}{\sqrt{y}}$.

$= \dfrac{\sqrt{25x^2}\sqrt{2x}}{\sqrt{9}}$ Factor the radicand. Use $\sqrt{xy} = \sqrt{x}\sqrt{y}$.

$= \dfrac{5x\sqrt{2x}}{3}$ Simplify the numerator; simplify the denominator.

b) $\sqrt{\dfrac{48}{16}}$

$= \dfrac{4\sqrt{3}}{4}$ Because $\dfrac{4}{4} = 1$, reduce.

$= \sqrt{3}$ This is the final answer.

However, in this example, you would simplify the fraction first:

$\sqrt{\dfrac{48}{16}} = \sqrt{3}$ Because $\dfrac{48}{16} = 3$.

c) $\dfrac{4\sqrt{54a^3}}{\sqrt{25}} = \dfrac{4\sqrt{9a^2}\sqrt{6a}}{\sqrt{25}}$ Factor the radicand.

$= \dfrac{4 \cdot 3a\sqrt{6a}}{5}$ Simplify the square roots.

$= \dfrac{12a\sqrt{6a}}{5}$ Multiply.

d) $\sqrt[3]{\dfrac{64x^5}{27}} = \dfrac{\sqrt[3]{64x^5}}{\sqrt[3]{27}}$ Use $\sqrt[3]{\dfrac{x}{y}} = \dfrac{\sqrt[3]{x}}{\sqrt[3]{y}}$.

$= \dfrac{\sqrt[3]{64x^3}\sqrt[3]{x^2}}{\sqrt[3]{27}}$ Factor the numerator into perfect cubes.

$= \dfrac{4x\sqrt[3]{x^2}}{3}$ Take the cube roots.

e) $\sqrt{\dfrac{75p^4}{16m^2}} = \dfrac{\sqrt{75p^4}}{\sqrt{16m^2}}$ Use $\sqrt{\dfrac{x}{y}} = \dfrac{\sqrt{x}}{\sqrt{y}}$.

$= \dfrac{\sqrt{25p^4}\sqrt{3}}{\sqrt{16m^2}}$ Factor the numerator into perfect squares.

$= \dfrac{5p^2\sqrt{3}}{4m}$ Take the square root of the numerator and the square root of the denominator.

8.3 ADDING AND SUBTRACTING RADICAL EXPRESSIONS

The next section describes when it is possible to add or subtract radicals, and then the process for performing the addition or subtraction.

ADDING SIMPLIFIED RADICALS

Only radicals with the same index and the same radicand (called **similar radicals**) can be added or subtracted. We can use the distributive property to allow us to add or subtract the numbers in front of the radicals and keep the radicands the same. Note that this process is the same process used to add $4x + 7x = (4 + 7)x = 11x$.

Exercise 8.7

Combine terms, if possible:

a) $4\sqrt{7} - 6\sqrt{7}$

b) $\sqrt[3]{9} + 4\sqrt[3]{9}$

c) $3\sqrt{2} - 5\sqrt{2} + 8\sqrt{2}$

d) $3\sqrt{2} + 4\sqrt{5}$

e) $\sqrt[3]{5} + \sqrt{5}$

Solution 8.7

a) $4\sqrt{7} - 6\sqrt{7} = (4-6)\sqrt{7}$ Use the distributive property.

$= -2\sqrt{7}$ Simplify inside the parentheses: $4 - 6 = -2$.

b) $\sqrt[3]{9} + 4\sqrt[3]{9} = 1\sqrt[3]{9} + 4\sqrt[3]{9}$ There is an understood 1 in front of $\sqrt[3]{9}$.

$= (1+4)\sqrt[3]{9}$ Use the distributive property.

$= 5\sqrt[3]{9}$ Simplify inside the parentheses: $1 + 4 = 5$.

c) $3\sqrt{2}-5\sqrt{2}+8\sqrt{2}$

$=(3-5+8)\sqrt{2}$ — Use the distributive property.

$=6\sqrt{2}$ — Simplify inside the parentheses: $3 - 5 + 8 = 6$.

d) $3\sqrt{2}+4\sqrt{5}$ — These terms cannot be combined. Both terms are square roots, but the radicands are not equal.

e) $\sqrt[3]{5}+\sqrt{5}$ — These terms cannot be combined. The radicands are equal (5), but the indices (3 and understood 2) are not equal.

ADDING RADICALS THAT ARE NOT SIMPLIFIED

If the radicals are not already simplified, you may need to simplify before (or after) combining similar radicals.

Exercise 8.8

Combine terms, if possible. Assume all variables represent positive real numbers:

a) $5\sqrt{2}+\sqrt{8}$

b) $3\sqrt{75}-2\sqrt{3}+2\sqrt{27}$

c) $2\sqrt{125}+\sqrt{20}-\sqrt{10}$

d) $\sqrt{12x^2}-7x\sqrt{3}$

e) $4\sqrt[3]{16a^2}+\sqrt[3]{54a^2}$

f) $\frac{1}{2}\sqrt{75}+\frac{1}{3}\sqrt{12}$

g) $6\sqrt{50a^3b^2}+a\sqrt{98ab^2}-4ab\sqrt{72a}$

Solution 8.8

a) $5\sqrt{2}+\sqrt{8}=5\sqrt{2}+\sqrt{4\cdot 2}$ — Simplify first.

$=5\sqrt{2}+2\sqrt{2}$ — $\sqrt{4\cdot 2}=\sqrt{4}\sqrt{2}=2\sqrt{2}$.

$=(5+2)\sqrt{2}$ — Use the distributive property.

$=7\sqrt{2}$ — Simplify inside the parentheses: $5 + 2 = 7$.

b) $3\sqrt{75}-2\sqrt{3}+2\sqrt{27}$

$= 3\sqrt{25}\sqrt{3}-2\sqrt{3}+2\sqrt{9}\sqrt{3}$ Factor each radicand.

$=3\cdot 5\sqrt{3}-2\sqrt{3}+2\cdot 3\sqrt{3}$ Simplify each square root.

$=15\sqrt{3}-2\sqrt{3}+6\sqrt{3}$ Multiply.

$=(15-2+6)\sqrt{3}$ Use the distributive property.

$=19\sqrt{3}$ Simplify inside the parentheses: $15 - 2 + 6 = 19$.

c) $2\sqrt{125}+\sqrt{20}-\sqrt{10}$

$= 2\sqrt{25}\sqrt{5}+\sqrt{4}\sqrt{5}-\sqrt{10}$ Simplify each square root.

$=2\cdot 5\sqrt{5}+2\sqrt{5}-\sqrt{10}$ $\sqrt{25}=5$; $\sqrt{4}=2$.

$=10\sqrt{5}+2\sqrt{5}-\sqrt{10}$ Multiply.

$=(10+2)\sqrt{5}-\sqrt{10}$ Use the distributive property.

$=12\sqrt{5}-\sqrt{10}$

Stop here, because the radicands (5 and 10) are not equal.

d) $\sqrt{12x^2}-7x\sqrt{3}=\sqrt{4x^2}\sqrt{3}-7x\sqrt{3}$ Simplify each square root.

$=2x\sqrt{3}-7x\sqrt{3}$ $\sqrt{4x^2}=2x$.

$=(2x-7x)\sqrt{3}$ Use the distributive property.

$=-5x\sqrt{3}$ Simplify inside the parentheses: $2x-7x=-5x$.

e) $4\sqrt[3]{16a^2}+\sqrt[3]{54a^2}=4\sqrt[3]{8}\sqrt[3]{2a^2}+\sqrt[3]{27}\sqrt[3]{2a^2}$ Factor each radicand.

$=4\cdot 2\sqrt[3]{2a^2}+3\sqrt[3]{2a^2}$ $\sqrt[3]{8}=2$; $\sqrt[3]{27}=3$.

$=8\sqrt[3]{2a^2}+3\sqrt[3]{2a^2}$ Multiply.

$=(8+3)\sqrt[3]{2a^2}$ Use the distributive property.

$=11\sqrt[3]{2a^2}$ Simplify inside the parentheses: $8 + 3 = 11$

f) $\frac{1}{2}\sqrt{75}+\frac{1}{3}\sqrt{12}$

$= \frac{1}{2}\sqrt{25}\sqrt{3}+\frac{1}{3}\sqrt{4}\sqrt{3}$ Factor each radicand.

$= \frac{1}{2}\cdot 5\sqrt{3}+\frac{1}{3}\cdot 2\sqrt{3}$ Simplify each square root.

$= \left(\frac{5}{2}+\frac{2}{3}\right)\sqrt{3}$ Use the distributive property.

$= \frac{19}{6}\sqrt{3}$ $\frac{5}{2}\cdot\frac{3}{3}+\frac{2}{3}\cdot\frac{2}{2}=\frac{15}{6}+\frac{4}{6}=\frac{19}{6}$.

g) $6\sqrt{50a^3b^2}+a\sqrt{98ab^2}-4ab\sqrt{72a}$

$= 6\sqrt{25a^2b^2}\sqrt{2a}+a\sqrt{49b^2}\sqrt{2a}-4ab\sqrt{36}\sqrt{2a}$ Factor each radicand.

$= 6\cdot 5ab\sqrt{2a}+a\cdot 7b\sqrt{2a}-4ab\cdot 6\sqrt{2a}$ Simplify each square root.

$= 30ab\sqrt{2a}+7ab\sqrt{2a}-24ab\sqrt{2a}$ Multiply.

$= (30ab+7ab-24ab)\sqrt{2a}$ Use the distributive property.

$= 13ab\sqrt{2a}$

8.4 MULTIPLYING RADICAL EXPRESSIONS

Previously we used the rule $\sqrt[n]{x\cdot y}=\sqrt[n]{x}\sqrt[n]{y}$ to simplify radicals. We can use the same rule now to multiply radicals. To multiply $(a\sqrt[n]{x})(b\sqrt[n]{y})$:

1. Multiply $a\cdot b\sqrt[n]{x\cdot y}$. Keep the index the same. That is, multiply the numbers in front of the radicals, multiply the numbers inside the radicals (the radicands), and keep the index the same.

2. Simplify $\sqrt[n]{xy}$, if possible.

Exercise 8.9

Multiply:

a) $(2\sqrt{5})(-3\sqrt{6})$

b) $3\sqrt{2}(\sqrt{8}+4\sqrt{10})$

c) $(\sqrt{3}+4)(\sqrt{3}+2)$

d) $(2\sqrt{x}-3)(4\sqrt{x}+1)$

Solution 8.9

a) $(2\sqrt{5})(-3\sqrt{6}) = (2\cdot -3)\sqrt{5\cdot 6}$ — $(a\sqrt[n]{x})(b\sqrt[n]{y}) = a\cdot b\sqrt[n]{x\cdot y}$.

$= -6\sqrt{30}$

Stop here, because $\sqrt{36}$ has no perfect square factors.

b) $3\sqrt{2}(\sqrt{8}+4\sqrt{10}) = (3\sqrt{2})(\sqrt{8})+(3\sqrt{2})(4\sqrt{10})$ — Use the distributive property.

$= (3\sqrt{2})(1\sqrt{8})+(3\sqrt{2})(4\sqrt{10})$ — $\sqrt{8} = 1\sqrt{8}$.

$= (3\cdot 1\sqrt{2\cdot 8})+(3\cdot 4\sqrt{2\cdot 10})$ — $(a\sqrt[n]{x})(b\sqrt[n]{y}) = a\cdot b\sqrt[n]{x\cdot y}$.

$= 3\sqrt{16}+12\sqrt{20}$ — Multiply.

$= 3\cdot 4+12\sqrt{4}\sqrt{5}$ — Factor each radicand.

$= 12+24\sqrt{5}$ — Multiply before adding.

$= 12+24\sqrt{5}$

Stop here: Only similar radicals can be added.
You can simplify $\sqrt{8} = 2\sqrt{2}$ first and arrive at the same answer:

$3\sqrt{2}(\sqrt{8}+4\sqrt{10}) = 3\sqrt{2}(2\sqrt{2}+4\sqrt{10})$ — Simplify $\sqrt{8} = 2\sqrt{2}$.

$= (3\sqrt{2})(2\sqrt{2})+(3\sqrt{2})(4\sqrt{10})$ — Use the distributive property.

$= (3\cdot 2\sqrt{2\cdot 2})+(3\cdot 4\sqrt{2\cdot 10})$ — $(a\sqrt[n]{x})(b\sqrt[n]{y}) = a\cdot b\sqrt[n]{x\cdot y}$

$= 6\sqrt{4}+12\sqrt{20}$ — Multiply.

$= 6 \cdot 2 + 12 \cdot 2\sqrt{5}$	Use $\sqrt[n]{x^n} = x$.
$= 12 + 24\sqrt{5}$	Multiply before adding. This is the same answer as previously shown.
c) $(\sqrt{3} + 4)(\sqrt{3} + 2)$	Use FOIL.
F: $\sqrt{3}\sqrt{3} = \sqrt{9} = 3$	Multiply the first two terms.
O: $\sqrt{3} \cdot 2 = 2\sqrt{3}$	Multiply the outer two terms.
I: $4 \cdot \sqrt{3} = 4\sqrt{3}$	Multiply the inner two terms.
L: $4 \cdot 2 = 8$	Multiply the last two terms.
$= 3 + 2\sqrt{3} + 4\sqrt{3} + 8$	Add similar radicals: $2\sqrt{3} + 4\sqrt{3} = (2+4)\sqrt{3} = 6\sqrt{3}$
$= 11 + 6\sqrt{3}$	Add similar terms: 3 + 8 = 11.
d) $(2\sqrt{x} - 3)(4\sqrt{x} + 1)$	Use FOIL.
F: $(2\sqrt{x})(4\sqrt{x})$	Multiply the first two terms.
$= (2 \cdot 4\sqrt{x \cdot x})$	$(a\sqrt[n]{x})(b\sqrt[n]{y}) = a \cdot b\sqrt[n]{x \cdot y}$.
$= 8\sqrt{x^2}$	Multiply.
$= 8x$	$\sqrt{x^2} = x$.
O: $(2\sqrt{x})(1)$	Multiply the outer two terms.
$= 2\sqrt{x}$	Use the identity property of 1.
I: $(-3)(4\sqrt{x})$	Multiply the inner two terms.
$= (-3 \cdot 4\sqrt{x})$	$(a\sqrt[n]{x})(b\sqrt[n]{y}) = a \cdot b\sqrt[n]{x \cdot y}$
$= -12\sqrt{x}$	Multiply.

L: $(-3)(+1)$ — Multiply the last two terms.

$= -3$

$= 8x + 2\sqrt{x} - 12\sqrt{x} - 3$ — Add similar radicals: $2\sqrt{x} - 12\sqrt{x} = (2-12)\sqrt{x} = -10\sqrt{x}$

$= 8x - 10\sqrt{x} - 3$

Stop here: There are no other similar terms to be added.

SPECIAL PRODUCTS

Recall from Chapter 5 that FOIL can be used to demonstrate the following rules:

$(a + b)^2 = a^2 + 2ab + b^2$

$(a - b)^2 = a^2 - 2ab + b^2$

$(a + b)(a - b) = a^2 - b^2$

When a and b are replaced with radicals, these rules still hold true.

Exercise 8.10

Multiply:

a) $(\sqrt{2}+6)^2$

b) $(5\sqrt{2}-\sqrt{3})^2$

c) $(4+\sqrt{3})(4-\sqrt{3})$

d) $(\sqrt{x}+7)^2$

e) $(\sqrt{y}-4)^2$

f) $(3+\sqrt{m})(3-\sqrt{m})$

Solution 8.10

a) $(\sqrt{2}+6)^2 = (\sqrt{2})^2 + 2(\sqrt{2})(6) + (6)^2$ — $(a + b)^2 = a^2 + 2ab + b^2$; $a = \sqrt{2}$ and $b = 6$.

$= 2 + 12\sqrt{2} + 36$ — $(\sqrt{2})^2 = 2$.

$= 38 + 12\sqrt{2}$ — Combine similar terms: $2 + 36 = 38$.

b) $(5\sqrt{2}-\sqrt{3})^2$

$= (5\sqrt{2})^2 - 2(5\sqrt{2})(\sqrt{3}) + (\sqrt{3})^2$ — $(a-b)^2 = a^2 - 2ab + b^2$; $a = 5\sqrt{2}$ and $b = \sqrt{3}$.

$= 25\sqrt{4} - 2(5\sqrt{6}) + 1\sqrt{9}$ — $(a\sqrt[n]{x})(b\sqrt[n]{y}) = a \cdot b\sqrt[n]{x \cdot y}$.

$= 25 \cdot 2 - 10\sqrt{6} + 1 \cdot 3$ — Simplify the square roots.

$= 50 - 10\sqrt{6} + 3$ — Multiply.

$= 53 - 10\sqrt{6}$ — Combine similar terms: $50 + 3 = 53$.

Stop here: Only similar radicals can be added.

c) $(4+\sqrt{3})(4-\sqrt{3}) = (4)^2 - (\sqrt{3})^2$ — $(a+b)(a-b) = a^2 - b^2$; $a = 4$ and $b = \sqrt{3}$

$= 16 - 3$ — $4^2 = 16$, $(\sqrt{3})^2 = 3$.

$= 13$

d) $(\sqrt{x}+7)^2 = (\sqrt{x})^2 + 2\sqrt{x}(7) + (7)^2$ — $(a+b)^2 = a^2 + 2ab + b^2$; $a = \sqrt{x}$ and $b = 7$.

$= x + 14\sqrt{x} + 49$ — $(\sqrt{x})^2 = x$.

Stop here: There are no similar radicals or similar terms to combine.

e) $(\sqrt{y}-4)^2 = (\sqrt{y})^2 - 2(\sqrt{y})(4) + (4)^2$ — $(a-b)^2 = a^2 - 2ab + b^2$; $a = \sqrt{y}$ and $b = 4$.

$= y - 8\sqrt{y} + 16$ — $(\sqrt{y})^2 = y$.

Stop here: There are no similar radicals or similar terms to combine.

f) $(3+\sqrt{m})(3-\sqrt{m}) = (3)^2 - (\sqrt{m})^2$ — $(a+b)(a-b) = a^2 - b^2$; $a = 3$ and $b = \sqrt{m}$.

$= 9 - m$ — $(\sqrt{m})^2 = m$.

MULTIPLYING CONJUGATES

The conjugate of a binomial $a - b$ is $a + b$, and the conjugate of $a + b$ is $a - b$. That is, you can find the conjugate of a binomial by changing the sign between the two terms.

Exercise 8.11

Find the conjugate of each of the following:

a) $5+\sqrt{3}$

b) $3-\sqrt{3}$

c) $\sqrt{7}+\sqrt{2}$

d) $2\sqrt{3}-4\sqrt{11}$

e) $-4+\sqrt{3}$

Solution 8.11

a) The conjugate of $5+\sqrt{3}$ is $5-\sqrt{3}$. Change the sign between the two terms.

b) The conjugate of $3-\sqrt{3}$ is $3+\sqrt{3}$. Change the sign between the two terms.

c) The conjugate of $\sqrt{7}+\sqrt{2}$ is $\sqrt{7}-\sqrt{2}$. Change the sign between the two terms.

d) The conjugate of $2\sqrt{3}-4\sqrt{11}$ is $2\sqrt{3}+4\sqrt{11}$ Change the sign between the two terms.

e) The conjugate of $-4+\sqrt{3}$ is $-4-\sqrt{3}$. Change the sign between the two terms.

The product of conjugates can be found using the special product formula:

$$(a + b)(a - b) = a^2 - b^2.$$

Exercise 8.12

Find the product of the following conjugates:

a) $(5+\sqrt{3})(5-\sqrt{3})$

b) $(3-\sqrt{3})(3+\sqrt{3})$

c) $(\sqrt{7}+\sqrt{2})(\sqrt{7}-\sqrt{2})$

d) $(2\sqrt{3}-4\sqrt{11})(2\sqrt{3}+4\sqrt{11})$

Solution 8.12

a) $(5+\sqrt{3})(5-\sqrt{3})=(5)^2-(\sqrt{3})^2$ Use $(a + b)(a - b) = a^2 - b^2$; $a=5$ and $b=\sqrt{3}$.

$=25-3$ Square both terms.

$=22$ Combine similar terms.

b) $(3-\sqrt{3})(3+\sqrt{3})=(3)^2-(\sqrt{3})^2$ — Use $(a+b)(a-b)=a^2-b^2$; $a=3$ and $b=\sqrt{3}$.

$=9-3$ — Square both terms.

$=6$ — Combine similar terms.

c) $(\sqrt{7}+\sqrt{2})(\sqrt{7}-\sqrt{2})=(\sqrt{7})^2-(\sqrt{2})^2$ — Use $(a+b)(a-b)=a^2-b^2$; $a=\sqrt{7}$ and $b=\sqrt{2}$.

$=7-2$ — Square both terms.

$=5$ — Combine similar terms.

d) $(2\sqrt{3}-4\sqrt{11})(2\sqrt{3}+4\sqrt{11})=(2\sqrt{3})^2-(4\sqrt{11})^2$ — Use $(a+b)(a-b)=a^2-b^2$; $a=2\sqrt{3}$ and $b=4\sqrt{11}$.

$=(2\cdot 2\sqrt{3\cdot 3})-(4\cdot 4\sqrt{11\cdot 11})$ — $(a\sqrt[n]{x})(b\sqrt[n]{y})=a\cdot b\sqrt[n]{x\cdot y}$

$=4\sqrt{9}-16\sqrt{121}$ — Multiply.

$=4\cdot 3-16\cdot 11$ — $\sqrt{9}=3$; $\sqrt{121}=11$.

$=12-176$ — Multiply before subtracting.

$=-164$ — Combine similar terms.

Look back at these products of conjugates. Notice that each conjugate involved one or two radicals. However, *none* of the products contained radicals! We make use of this concept in the next section.

8.5 DIVIDING RADICAL EXPRESSIONS

We have already looked at division problems like $\sqrt{\dfrac{16}{25}}=\dfrac{4}{5}$ and $\dfrac{\sqrt{72x^3}}{\sqrt{49}}=\dfrac{6x\sqrt{2x}}{7}$, where the denominators became rational numbers. Other division problems may require the technique called rationalizing the denominator, which is outlined below.

To divide with radicals:

1. Reduce, if possible, using $\frac{\sqrt{a}}{\sqrt{b}} = \sqrt{\frac{a}{b}}$.
2. If the denominator is a monomial, multiply by $\frac{\sqrt{b}}{\sqrt{b}}$.
3. If the denominator is a binomial, multiply the numerator and denominator by the conjugate of the denominator.
4. Reduce, if possible.

RATIONALIZING MONOMIAL DENOMINATORS

Notice from the problems that follow that a monomial denominator may begin as a single term in a denominator that contains a radical or may result from simplifying a fraction under a radical.

Exercise 8.13

Simplify:

a) $\sqrt{\frac{1}{5}}$

b) $\sqrt{\frac{20}{7}}$

c) $\frac{10\sqrt{12}}{2\sqrt{6}}$

d) $\sqrt[3]{\frac{1}{2}}$

e) $\sqrt[3]{\frac{2}{25}}$

Solution 8.13

a) $\sqrt{\frac{1}{5}} = \frac{\sqrt{1}}{\sqrt{5}}$

The fraction will not reduce. Rewrite the fraction using $\sqrt{\frac{a}{b}} = \frac{\sqrt{a}}{\sqrt{b}}$.

$=\sqrt{\frac{1}{5}}\cdot\frac{\sqrt{5}}{\sqrt{5}}$ Multiply by $\frac{\sqrt{5}}{\sqrt{5}}$ to make the denominator a rational number.

$=\frac{\sqrt{5}}{\sqrt{25}}$ Use the rule for multiplying radicals: $(a\sqrt[n]{x})(b\sqrt[n]{y})=a\cdot b\sqrt[n]{x\cdot y}$.

$=\frac{\sqrt{5}}{5}$ Simplify the denominator: $\sqrt{25}=5$.

b) $\sqrt{\frac{20}{7}}=\frac{\sqrt{20}}{\sqrt{7}}$ The fraction will not reduce. Rewrite the fraction using $\sqrt{\frac{a}{b}}=\frac{\sqrt{a}}{\sqrt{b}}$.

$=\frac{\sqrt{20}}{\sqrt{7}}\cdot\frac{\sqrt{7}}{\sqrt{7}}$ Multiply by $\frac{\sqrt{7}}{\sqrt{7}}$ to make the denominator a rational number.

$=\frac{\sqrt{140}}{\sqrt{49}}$ Use the rule for multiplying radicals: $(a\sqrt[n]{x})(b\sqrt[n]{y})=a\cdot b\sqrt[n]{x\cdot y}$.

$=\frac{\sqrt{4}\sqrt{35}}{7}$ Simplify the numerator and the denominator.

$=\frac{2\sqrt{35}}{7}$ This is the final answer.

c) $\frac{10\sqrt{12}}{2\sqrt{6}}=\frac{10}{2}\sqrt{\frac{12}{6}}$ Use $\frac{\sqrt{a}}{\sqrt{b}}=\sqrt{\frac{a}{b}}$.

$=5\sqrt{2}$ Reduce.

There is no radical in the denominator, so we are finished.

d) $\sqrt[3]{\dfrac{1}{2}} = \dfrac{\sqrt[3]{1}}{\sqrt[3]{2}}$ The fraction will not reduce. Rewrite the fraction using $\sqrt[3]{\dfrac{a}{b}} = \dfrac{\sqrt[3]{a}}{\sqrt[3]{b}}$.

$= \dfrac{\sqrt[3]{1}}{\sqrt[3]{2}} \cdot \dfrac{\sqrt[3]{4}}{\sqrt[3]{4}}$ This is not a square root. You must multiply by $\sqrt[3]{4}$ to produce a rational denominator.

$= \dfrac{\sqrt[3]{4}}{\sqrt[3]{8}}$ Use the rule for multiplying radicals: $(a\sqrt[n]{x})(b\sqrt[n]{y}) = a \cdot b\sqrt[n]{x \cdot y}$.

$= \dfrac{\sqrt[3]{4}}{2}$ Simplify the denominator.

e) $\sqrt[3]{\dfrac{2}{25}} = \dfrac{\sqrt[3]{2}}{\sqrt[3]{25}}$ The fraction will not reduce. Rewrite the fraction using $\sqrt[3]{\dfrac{a}{b}} = \dfrac{\sqrt[3]{a}}{\sqrt[3]{b}}$.

$= \dfrac{\sqrt[3]{2}}{\sqrt[3]{25}} \cdot \dfrac{\sqrt[3]{5}}{\sqrt[3]{5}}$ To produce a perfect cube, we must produce $\sqrt[3]{125}$, so multiply by $\sqrt[3]{5}$.

$= \dfrac{\sqrt[3]{2 \cdot 5}}{\sqrt[3]{125}}$ Multiply the numerators; multiply the denominators.

$= \dfrac{\sqrt[3]{10}}{5}$ Simplify the denominator.

RATIONALIZING BINOMIAL DENOMINATORS

When a denominator is a binomial (two terms) that involves one or two radicals, we must multiply the numerator and denominator by the conjugate of the denominator to finish with a denominator that does not contain a radical.

Exercise 8.14

Rationalize the denominator by multiplying by the conjugate in each of the following:

a) $\dfrac{\sqrt{2}}{5 + \sqrt{3}}$

b) $\dfrac{12}{3-\sqrt{3}}$

c) $\dfrac{2-\sqrt{5}}{\sqrt{7}+\sqrt{2}}$

Solution 8.14

a) $\dfrac{\sqrt{2}}{5+\sqrt{3}}=\dfrac{\sqrt{2}}{5+\sqrt{3}}\cdot\dfrac{(5-\sqrt{3})}{(5-\sqrt{3})}$ The conjugate of $5+\sqrt{3}$ is $5-\sqrt{3}$.

$=\dfrac{5\sqrt{2}-\sqrt{6}}{(5)^2-(\sqrt{3})^2}$ Use the distributive property to multiply the numerator. Use $(a+b)(a-b)=a^2-b^2$ to multiply the conjugates in the denominator.

$=\dfrac{5\sqrt{2}-\sqrt{6}}{25-3}$ Square both terms in the denominator.

$=\dfrac{5\sqrt{2}-\sqrt{6}}{22}$ Combine similar terms in the denominator.

b) $\dfrac{12}{3-\sqrt{3}}=\dfrac{12}{3-\sqrt{3}}\cdot\dfrac{(3+\sqrt{3})}{(3+\sqrt{3})}$ The conjugate of $3-\sqrt{3}$ is $3+\sqrt{3}$.

$=\dfrac{12(3+\sqrt{3})}{(3)^2-(\sqrt{3})^2}$ Use $(a+b)(a-b)=a^2-b^2$ to multiply the conjugates in the denominator. Leave the numerator factored.

$=\dfrac{12(3+\sqrt{3})}{9-3}$ Square both terms in the denominator.

$=\dfrac{12(3+\sqrt{3})}{6}$ Reduce $\dfrac{12}{6}=2$.

$=2(3+\sqrt{3})$

Sometimes leaving the numerator factored can make reducing easier. If we had multiplied $12(3+\sqrt{3})=36+12\sqrt{3}$, we would have had to factor before reducing.

c) $\frac{2-\sqrt{5}}{\sqrt{7}+\sqrt{2}}=\frac{2-\sqrt{5}}{\sqrt{7}+\sqrt{2}}\cdot\frac{(\sqrt{7}-\sqrt{2})}{(\sqrt{7}-\sqrt{2})}$ The conjugate of $\sqrt{7}+\sqrt{2}$ is $\sqrt{7}-\sqrt{2}$.

$=\frac{2\sqrt{7}-2\sqrt{2}-\sqrt{5}\sqrt{7}+\sqrt{5}\sqrt{2}}{(\sqrt{7})^2-(\sqrt{2})^2}$ Use FOIL to multiply the numerators. Use $(a+b)(a-b)=a^2-b^2$ to multiply the conjugates in the denominator.

$=\frac{2\sqrt{7}-2\sqrt{2}-\sqrt{35}+\sqrt{10}}{7-2}$ Square both terms in the denominator.

$=\frac{2\sqrt{7}-2\sqrt{2}-\sqrt{35}+\sqrt{10}}{5}$ There are no common factors in the numerator and denominator, so the fraction does not reduce.

8.6 SOLVING EQUATIONS CONTAINING RADICAL EXPRESSIONS

Solving equations containing radicals makes use of the technique of squaring both sides of an equation. This technique can produce extraneous solutions; that is, solutions that will not check in the original equation. For these types of problems you must check every solution. To solve an equation containing radicals, do the following:

1. Isolate one radical on one side of the equation.
2. Square both sides of the equation. Remember that $(a+b)^2=a^2+2ab+b^2$.
3. Combine similar terms.
4. If a radical remains in the equation, go back to Step 1.
5. Solve the resulting equation. If the resulting equation is linear, isolate the variable. If the resulting equation is a quadratic equation, get 0 on one side and factor.
6. Check all proposed solutions in the original equation.

Exercise 8.15

Solve:

a) $\sqrt{x-2}=5$

b) $\sqrt{a+2}=-3$

c) $\sqrt{2p-3}+4=6$

d) $\sqrt{3x+1}=x-3$

e) $\sqrt{32+x}-\sqrt{x}=4$

Solution 8.15

a) $\sqrt{x-2}=5$

$(\sqrt{x-2})^2=(5)^2$	Square both sides.
$x-2=25$	$(\sqrt{x-2})^2=x-2$.
$x-2+2=25+2$	Add 2 to both sides of the equation.
$x=27$	This is the proposed solution.
Check:	
$\sqrt{27-2}=5$	Substitute $x=27$ into the original equation.
$\sqrt{25}=5$	This is true, so $x=27$ is the solution.

b) $\sqrt{a+2}=-3$

$(\sqrt{a+2})^2=(-3)^2$	Square both sides.
$a+2=9$	This is a linear equation. Isolate the variable.
$a+2-2=9-2$	Subtract 2 from both sides of the equation.
$a=7$	This is the proposed solution.
Check:	
$\sqrt{7+2}=-3$	Substitute $a=7$ into the original equation.
$\sqrt{9}=-3$	

False, so there is no solution. Remember that if there is no sign in front of the radical, we are asking for the positive square root. Thus, $\sqrt{9}\neq-3$.

c) $\sqrt{2p-3}+4=6$	Isolate the radical.
$\sqrt{2p-3}+4-4=6-4$	Subtract 4 from both sides of the equation.
$\sqrt{2p-3}=2$	
$(\sqrt{2p-3})^2=(2)^2$	Square both sides.
$2p-3=4$	This is a linear equation. Isolate the variable.
$2p-3+3=4+3$	Add 3 to both sides of the equation.
$\frac{2p}{2}=\frac{7}{2}$	Multiply both sides by $\frac{1}{2}$.
$p=\frac{7}{2}$	This is the proposed solution.

Check:

$\sqrt{2\left(\frac{7}{2}\right)-3}+4=6$ Substitute $p=\frac{7}{2}$ into the original equation.

$\sqrt{7-3}+4=6$

$\sqrt{4}+4=6$

$2+4=6$ This is true, so $p=\frac{7}{2}$ is the solution.

d) $\sqrt{3x+1}=x-3$

$(\sqrt{3x+1})^2=(x-3)^2$ Square both sides.

$3x+1=x^2-6x+9$ $(x-3)^2=(x)^2-2(x)(3)+(3)^2$.

$3x+1-3x-1=x^2-6x+9-3x-1$ This is a quadratic equation. Get 0 on one side of the equation.

$x^2-9x+8=0$

$(x-8)(x-1)=0$ Factor. Set each factor equal to 0 and solve.

$x=8$ or $x=1$ These are the proposed solutions.

Check:

If $x=8$:

$\sqrt{3(8)+1}=8-3$

$\sqrt{25}=8-3$

$5=5$ This is true.

If $x=1$:

$\sqrt{3(1)+1}=1-3$

$2=-2$ This is false.

Only $x=8$ is a solution.

e) $\sqrt{32+x}-\sqrt{x}=4$ Isolate one radical.

$\sqrt{32+x}-\sqrt{x}=4$ Add $\sqrt{x}$ to both sides.

$(\sqrt{32+x})^2=(\sqrt{x}+4)^2$ Square both sides.

$32+x=x+8\sqrt{x}+16$ $(\sqrt{x}+4)^2=(\sqrt{x})^2+2\sqrt{x}(4)+(4)^2$

$32+x-x-16=x+8\sqrt{x}+16-x-16$	A radical remains in the equation; isolate the radical.
$16=8\sqrt{x}$	Simplify.
$\frac{1}{8}(16)=\frac{1}{8}(8\sqrt{x})$	Multiply both sides by $\frac{1}{8}$.
$(2)^2=(\sqrt{x})^2$	Square both sides.
$4=x$	This is the proposed solution.

Check:

$\sqrt{32+4}-\sqrt{4}=4$

$\sqrt{36}-\sqrt{4}=4$

$6-2=4$ This is true, so $x=4$ is the solution.

TEST YOURSELF

1) Find each square root:

(a) $\sqrt{64}$

(b) $-\sqrt{81}$

(c) $\sqrt{\frac{9}{4}}$

(d) $\frac{-\sqrt{49}}{-\sqrt{4}}$

2) Find each root, if it exists:

(a) $\sqrt[3]{-27}$

(b) $\sqrt{1}$

(c) $\sqrt[3]{343}$

(d) $\sqrt{-16}$

3) Simplify:

(a) $\sqrt{147x^3y^2}$

(b) $5a\sqrt{50a^2b}$

(c) $p^2q\sqrt[3]{64p^4q^5}$

4) Simplify:

(a) $\sqrt{\frac{9}{100}}$

(b) $\sqrt[3]{\frac{216x^6}{343y^3}}$

(c) $\sqrt{\frac{32}{50}}$

5) Simplify:

(a) $\sqrt{\frac{98x}{25}}$

(b) $\sqrt{\frac{360}{10}}$

(c) $\sqrt[3]{\frac{125x^7}{8}}$

6) Perform the indicated operations:

(a) $\sqrt{13}+4\sqrt{13}$

(b) $5\sqrt{72}+\sqrt{8}$

(c) $\sqrt{200t}-\sqrt{128t}+6\sqrt{2t}$

(d) $4\sqrt{147}-\sqrt{27}$

7) Multiply:

(a) $(4\sqrt{3})(-5\sqrt{7})$

(b) $\sqrt{10}(3-4\sqrt{6})$

(c) $(3-2\sqrt{5})(4-6\sqrt{5})$

(d) $(3\sqrt{x}+1)(2\sqrt{x}-4)$

8) Multiply:

(a) $(3+\sqrt{5})^2$

(b) $(2\sqrt{3}-4)^2$

(c) $(5-\sqrt{2})(5+\sqrt{2})$

(d) $(\sqrt{a}+4)^2$

9) Simplify:

(a) $\sqrt{\frac{11}{3}}$

(b) $\sqrt{\frac{27}{7}}$

(c) $\frac{8\sqrt{16}}{2\sqrt{8}}$

(d) $\sqrt[3]{\frac{1}{3}}$

10) Rationalize each denominator:

(a) $\dfrac{5}{2+\sqrt{3}}$

(b) $\dfrac{\sqrt{6}}{3-\sqrt{5}}$

(c) $\dfrac{4+\sqrt{3}}{\sqrt{2}+\sqrt{3}}$

11) Solve. Be careful to check each proposed solution in the original equation:

(a) $\sqrt{3x+1}=4$

(b) $\sqrt{x-4}=-2$

(c) $\sqrt{4a+1}-3=2$

(d) $\sqrt{x+3}=x-3$

(e) $4-\sqrt{x}=\sqrt{x+8}$

TEST YOURSELF ANSWERS

1) (a) 8

(b) -9

(c) $\frac{3}{2}$

(d) $\frac{7}{2}$

2) (a) -3

(b) 1

(c) 7

(d) Does not exist

3) (a) $7xy\sqrt{3x}$

(b) $25a^2\sqrt{2b}$

(c) $4p^3q^2\sqrt[3]{pq^2}$

4) (a) $\frac{3}{10}$

(b) $\frac{6x^2}{7y}$

(c) $\frac{4}{5}$

5) (a) $\frac{7\sqrt{2x}}{5}$

(b) 6

(c) $\frac{5x^2\sqrt[3]{x}}{2}$

6) (a) $5\sqrt{13}$

(b) $32\sqrt{2}$

(c) $8\sqrt{2t}$

(d) $25\sqrt{3}$

7) (a) $-20\sqrt{21}$

(b) $3\sqrt{10}-8\sqrt{15}$

(c) $72-26\sqrt{5}$

(d) $6x-10\sqrt{x}-4$

8) (a) $14+6\sqrt{5}$

(b) $28-16\sqrt{3}$

(c) 23

(d) $a+8\sqrt{a}+16$

9) (a) $\frac{\sqrt{33}}{3}$

(b) $\frac{3\sqrt{21}}{7}$

(c) $4\sqrt{2}$

(d) $\frac{\sqrt[3]{9}}{3}$

10) (a) $10-5\sqrt{3}$

(b) $\dfrac{3\sqrt{6}+\sqrt{30}}{4}$

(c) $-4\sqrt{2}+4\sqrt{3}-\sqrt{6}+3$

11) (a) 5

(b) No solution

(c) 6

(d) 6

(e) 1

CHAPTER 9

Quadratic Equations

Quadratic equations were first introduced in Chapter 6, where solutions were obtained by factoring and setting each factor equal to 0. In this chapter, we expand our techniques for solving quadratic equations.

9.1 SOLVING QUADRATIC EQUATIONS USING THE SQUARE ROOT METHOD

A quadratic equation can be written as $ax^2 + bx + c = 0$, where a, b, and c are real numbers and $a \neq 0$. A quadratic equation like $x^2 = 16$ can be solved mentally to give solutions of $x = 4$ or $x = -4$. In so doing, you are making use of the **square root property,** which is stated as follows:

If $a^2 = b$, then $a = +\sqrt{b}$ or $a = -\sqrt{b}$.

***Note*:** If $b \geq 0$, the solutions are real numbers. If $b < 0$, the solutions are complex numbers (see the "Complex Numbers" section later in this chapter).

To solve using the square root property:

1. Isolate the squared term, which may be x^2 or $(x + d)^2$ or $(x - d)^2$, on one side of the equation and isolate the constant term on the other side of the equation.
2. Take the square root of both sides of the equation.
3. Use $+\sqrt{\text{constant}}$ and $-\sqrt{\text{constant}}$, which is usually written $\pm\sqrt{\text{constant}}$.

4. Simplify the radicals.
5. Solve for x, if necessary.
6. Check the answers.

Exercise 9.1

Solve:

a) $x^2 = 25$
b) $y^2 = 6$
c) $(x - 2)^2 = 16$
d) $(x + 1)^2 = 8$
e) $2m^2 - 7 = 47$
f) $(x - 3)^2 = -4$
g) $(4x + 1)^2 = 25$
h) $(3y - 4)^2 = 16$
i) $(2x - 1)^2 = 12$

Solution 9.1

a) $x^2 = 25$

$\sqrt{x^2} = \pm\sqrt{25}$ — Take the square root of both sides, using $\pm$ on the right.

$x = \pm 5$ — Simplify the radicals.

This means $x = 5$ or $x = -5$.

b) $y^2 = 6$

$\sqrt{y^2} = \pm\sqrt{6}$ — Take the square root of both sides, using $\pm$ on the right.

$y = \pm\sqrt{6}$ — Simplify the radicals. The radical on the right side is already simplified.

c) $(x - 2)^2 = 16$

$\sqrt{(x-2)^2} = \pm\sqrt{16}$ — Take the square root of both sides, using $\pm$ on the right.

$x - 2 = \pm 4$ — Simplify the radicals.

$x - 2 + 2 = 2 \pm 4$ — Solve for x by adding 2 to both sides.

$x = 2 \pm 4$

Because 2 can be combined with 4, we separate the two possible solutions to solve for x as follows:

$x - 2 = 4$ or $x - 2 = -4$ — Solve for x.

$x - 2 + 2 = 4 + 2$ or

$x - 2 + 2 = -4 + 2$ Add 2 to both sides.

$x = 6$ or $x = -2$

d) $(x + 1)^2 = 8$

$\sqrt{(x+1)^2} = \pm\sqrt{8} = \pm\sqrt{4 \cdot 2}$ Take the square root of both sides, using ± on the right.

$x+1=\pm 2\sqrt{2}$ Simplify the radicals.

$x+1-1=-1\pm 2\sqrt{2}$ Solve for x. Subtract 1 from both sides of the equation.

Because -1 cannot be combined with $\pm 2\sqrt{2}$, we do not need to separate the two solutions to solve for x. Note that we usually write the answer in the form shown, rather than $\pm 2\sqrt{2}-1$.

e) $2m^2 - 7 = 47$

$2m^2 - 7 + 7 = 47 + 7$ Isolate the squared term first by adding 7 to both sides.

$2m^2 = 54$

$\frac{2m^2}{2} = \frac{54}{2}$ Divide both sides by 2.

$m^2 = 27$

$\sqrt{m^2} = \pm\sqrt{27}$ Take the square root of both sides, using ± on the right.

$m = \pm 3\sqrt{3}$ Simplify the radicals: $\sqrt{27} = \sqrt{9}\sqrt{3} = 3\sqrt{3}$

f) $(x - 3)^2 = -4$

$\sqrt{(x-3)^2} = \pm\sqrt{-4}$ Take the square root of both sides, using ± on the right.

These will *not* be real solutions, because the square root of a negative number is not a real number. (This type of equation will be solved in the "Complex Numbers" section later in this chapter.) There are no real number solutions.

g) $(4x + 1)^2 = 25$

$\sqrt{(4x+1)^2} = \pm\sqrt{25}$ Take the square root of both sides, using ± on the right.

$4x + 1 = \pm 5$ Simplify the radicals.

$4x + 1 - 1 = \pm 5 - 1$ Isolate $4x$ by subtracting 1 from both sides.

$\frac{4x}{4} = \frac{-1 \pm 5}{4}$ Divide both sides by 4.

$x = \frac{-1+5}{4}$ or $x = \frac{-1-5}{4}$ Separate the two solutions.

$x = \frac{4}{4} = 1$ or $x = \frac{-6}{4} = -\frac{3}{2}$ Simplify each solution.

h) $(3y^2 - 4)^2 = 16$

$\sqrt{(3y-4)^2} = \pm\sqrt{16}$ Take the square root of both sides, using $\pm$ on the right.

$3y - 4 = \pm 4$ Simplify the radicals.

$3y - 4 + 4 = \pm 4 + 4$ Isolate $3y$ by adding 4 to both sides.

$\frac{3y}{3} = \frac{4 \pm 4}{3}$ Divide both sides by 3.

$y = \frac{4+4}{3}$ or $y = \frac{4-4}{3}$ Separate the two solutions.

$y = \frac{8}{3}$ or $y = \frac{0}{3} = 0$ Simplify each solution.

i) $(2x - 1)^2 = 12$

$\sqrt{(2x-1)^2} = \pm\sqrt{12}$ Take the square root of both sides, using $\pm$ on the right.

$2x - 1 = \pm 2\sqrt{3}$ Simplify each radical: $\sqrt{12} = \sqrt{4}\sqrt{3} = 2\sqrt{3}$.

$2x - 1 + 1 = 1 \pm 2\sqrt{3}$ Isolate $2x$ by adding 1 to both sides.

$\frac{2x}{2} = \frac{1 \pm 2\sqrt{3}}{2}$ Divide both sides by 2.

$x = \frac{1 \pm 2\sqrt{3}}{2}$ The right side does not simplify further.

9.2 SOLVING QUADRATIC EQUATIONS USING COMPLETING THE SQUARE

We will use the technique of completing the square in this section to solve quadratic equations. When you continue your study of mathematics, you will also use this process to find such things as the center of a circle and the vertex of a parabola.

To solve a quadratic equation by completing the square, do the following:

1. Isolate the x^2 and x terms on one side of the equation so that the equation has the form $ax^2 + bx = c$.
2. If $a \neq 1$, divide every term by a, the coefficient of x^2. The equation should now have the form $x^2 + \frac{b}{a}x = \frac{c}{a}$.
3. Add $\left(\frac{1}{2}\right)$ times the coefficient of x to both sides of the equation.
4. Simplify the right side of the equation.
5. Write the left side as a perfect square trinomial.
6. Solve using the square root property.

Exercise 9.2

Solve by completing the square:

a) $x^2 + 6x - 7 = 0$

b) $x^2 - 5x + 4 = 0$

c) $x^2 + 4x + 1 = 0$

Solution 9.2

a) $x^2 + 6x - 7 = 0$

$x^2 + 6x - 7 + 7 = 0 + 7$	Isolate $x^2 + 6x$ by adding 7 to both sides of the equation.
$x^2 + 6x = 7$	The coefficient of x^2 equals 1, so there is no need to divide.
$x^2 + 6x + 9 = 7 + 9$	$\left(\frac{1}{2}\cdot 6\right)^2 = (3)^2 = 9$. Add 9 to both sides of the equation.
$x^2 + 6x + 9 = 16$	Simplify the right side of the equation.
$(x + 3)^2 = 16$	Factor the left side.
$\sqrt{(x+3)^2} = \pm\sqrt{16}$	Take the square root of both sides.
$x + 3 = \pm 4$	Simplify the radicals.
$x + 3 = 4$ or $x + 3 = -4$	Solve for x.
$x + 3 - 3 = 4 - 3$ or $x + 3 - 3 = -4 - 3$	
$x = 1$ or $x = -7$	

b) $x^2 - 5x + 4 = 0$

$x^2 - 5x = -4$ — Isolate $x^2 - 5x$ by subtracting 4 from both sides of the equation. The coefficient of x^2 is 1, so there is no need to divide.

$x^2 - 5x + \frac{25}{4} = -4 + \frac{25}{4}$ — $\left(\frac{1}{2} \cdot -5\right)^2 = \left(-\frac{5}{2}\right)^2 = \frac{25}{4}$. Add $\frac{25}{4}$ to both sides of the equation.

$x^2 - 5x + \frac{25}{4} = \frac{9}{4}$ — Simplify the right side of the equation: $-\frac{4}{1} + \frac{25}{4} = -\frac{16}{4} + \frac{25}{4} = \frac{9}{4}$.

$\left(x - \frac{5}{2}\right)^2 = \frac{9}{4}$ — Factor $x^2 - 5x + \frac{25}{4}$.

$\sqrt{\left(x - \frac{5}{2}\right)^2} = \pm\sqrt{\frac{9}{4}}$ — Take the square root of both sides.

$x - \frac{5}{2} = \pm\frac{3}{2}$ — Simplify the radicals.

$x - \frac{5}{2} = \frac{3}{2}$ or $x - \frac{5}{2} = -\frac{3}{2}$ — Separate the solutions. Solve for x.

$$x - \frac{5}{2} + \frac{5}{2} = \frac{3}{2} + \frac{5}{2} \text{ or } x - \frac{5}{2} + \frac{5}{2} = -\frac{3}{2} + \frac{5}{2}$$

$$x = \frac{8}{2} \text{ or } x = \frac{2}{2}$$

$$x = 4 \text{ or } x = 1$$

c) $x^2 + 4x + 1 = 0$

$x^2 + 4x = -1$ — Isolate $x^2 + 4x$ by subtracting 1 from both sides. The coefficient of x^2 is 1, so there is no need to divide.

$x^2 + 4x + 4 = -1 + 4$ — $\left(\frac{1}{2} \cdot 4\right)^2 = (2)^2 = 4$ so add 4 to both sides of the equation.

$x^2 + 4x + 4 = 3$	Simplify the right side.
$(x + 2)^2 = 3$	Factor $x^2 + 4x + 4$.
$\sqrt{(x+2)^2} = \pm\sqrt{3}$	Take the square root of both sides.
$x+2 = \pm\sqrt{3}$	Simplify the radicals.
$x+2-2 = -2\pm\sqrt{3}$	Subtract 2 from both sides.
$x = -2+\sqrt{3}$ or $x = -2-\sqrt{3}$	Solve for x. The answer may be written using $x = -2\pm\sqrt{3}$, or as two separate answers.

Exercise 9.3

Solve by completing the square:

a) $2x^2 - 4x - 6 = 0$

b) $2x^2 - 10x + 12 = 0$

c) $4x^2 + 9x + 3 = 0$

d) $3x^2 + 5x + 3 = 0$

Solution 9.3

a) $2x^2 - 4x - 6 = 0$

$2x^2 - 4x - 6 + 6 = 0 + 6$	Isolate $2x^2 - 4x$ by adding 6 to both sides of the equation.
$2x^2 - 4x = 6$	
$\frac{2x^2}{2} - \frac{4x}{2} = \frac{6}{2}$	Divide every term by 2 to make the coefficient of x^2 equal 1.
$x^2 - 2x = 3$	
$x^2 - 2x + 1 = 3 + 1$	$\left(\frac{1}{2}\cdot -2\right)^2 = (-1)^2 = 1$, so add 1 to both sides of the equation.
$x^2 - 2x + 1 = 4$	Simplify the right side.
$(x - 1)^2 = 4$	Factor $x^2 - 2x + 1$.
$\sqrt{(x-1)^2} = \pm\sqrt{4}$	Take the square root of both sides.
$x-1 = \pm 2$	Simplify the radicals.
$x-1+1 = 1\pm 2$	Solve for x by adding 1 to both sides of the equation.

$x = 1 \pm 2$

$x = 1 + 2$ or $x = 1 - 2$ Separate into two equations to solve for x.

$x = 3$ or $x = -1$

b) $2x^2 - 10x + 12 = 0$

$2x^2 - 10x + 12 - 12 = 0 - 12$ Isolate $2x^2 - 10x$ by subtracting 12 from both sides of the equation.

$2x^2 - 10x = -12$

$\frac{2x^2}{2} - \frac{10x}{2} = \frac{-12}{2}$ Divide every term by 2 to make the coefficient of x^2 equal 1.

$x^2 - 5x = -6$

$x^2 - 5x + \frac{25}{4} = -6 + \frac{25}{4}$ $\left(\frac{1}{2} \cdot -5\right)^2 = \frac{25}{4}$, so add $\frac{25}{4}$ to both sides of the equation.

$x^2 - 5x + \frac{25}{4} = \frac{1}{4}$ Simplify the right side: $-\frac{6}{1} \cdot \frac{4}{4} + \frac{25}{4} = \frac{-24}{4} + \frac{25}{4} = \frac{1}{4}$.

$\left(x - \frac{5}{2}\right)^2 = \frac{1}{4}$ Factor the left side.

$\sqrt{\left(x - \frac{5}{2}\right)^2} = \pm\sqrt{\frac{1}{4}}$ Take the square root of both sides.

$x - \frac{5}{2} = \pm\frac{1}{2}$ Simplify the radicals.

$x - \frac{5}{2} = \frac{1}{2}$ or $x - \frac{5}{2} = -\frac{1}{2}$

$x - \frac{5}{2} + \frac{5}{2} = \frac{1}{2} + \frac{5}{2}$ or $x - \frac{5}{2} + \frac{5}{2} = -\frac{1}{2} + \frac{5}{2}$

$x = \frac{6}{2}$ or $x = \frac{4}{2}$

$x = 3$ or $x = 2$

c) $4x^2 + 9x + 3 = 0$

$4x^2 + 9x + 3 - 3 = 0 - 3$ Isolate $4x^2 + 9x$ by subtracting 3 from both sides.

$4x^2 + 9x = -3$

$$\frac{4x^2}{4} + \frac{9x}{4} = -\frac{3}{4}$$

Divide every term by 4.

$$x^2 + \frac{9x}{4} = -\frac{3}{4}$$

$$x^2 + \frac{9x}{4} + \frac{81}{64} = -\frac{3}{4} + \frac{81}{64}$$

$\left(\frac{1}{2}\cdot\frac{9}{4}\right)^2 = \left(\frac{9}{8}\right)^2 = \frac{81}{64}$, so add $\frac{81}{64}$ to both sides of the equation.

$$x + \frac{9x}{4} + \frac{81}{64} = \frac{33}{64}$$

Simplify the right side of the equation:

$$\left(-\frac{3}{4}\cdot\frac{16}{16}\right) + \frac{81}{64} = -\frac{48}{64} + \frac{81}{64} = \frac{33}{64}$$

$$\left(x + \frac{9}{8}\right)^2 = \frac{33}{64}$$

Factor the left side.

$$\sqrt{\left(x + \frac{9}{8}\right)^2} = \pm\sqrt{\frac{33}{64}}$$

Take the square root of both sides.

$$x + \frac{9}{8} = \pm\frac{\sqrt{33}}{8}$$

Simplify the radicals.

$$x + \frac{9}{8} - \frac{9}{8} = -\frac{9}{8} \pm \frac{\sqrt{33}}{8}$$

Solve for x by subtracting $\frac{9}{8}$ from both sides of the equation.

$$x = -\frac{9}{8} \pm \frac{\sqrt{33}}{8}$$

d) $3x^2 + 5x + 3 = 0$

$3x^2 + 5x + 3 - 3 = 0 - 3$ Isolate $3x^2 + 5x$ by subtracting 3 from both sides of the equation.

$3x^2 + 5x = -3$

$$\frac{3x^2}{3}+\frac{5x}{3}=-\frac{3}{3}$$ Divide every term by 3.

$$x^2+\frac{5}{3}x=-1$$

$$x^2+\frac{5}{3}x+\frac{25}{36}=-1+\frac{25}{36}$$ $\left(\frac{1}{2}\cdot\frac{5}{3}\right)^2=\left(\frac{5}{6}\right)^2=\frac{25}{36}$, so add $\frac{25}{36}$ to both sides of the equation.

$$x^2+\frac{5}{3}x+\frac{25}{36}=-\frac{11}{36}$$ Simplify the right side: $-1+\frac{25}{36}=-\frac{36}{36}+\frac{25}{36}=-\frac{11}{36}$.

$$\left(x+\frac{5}{6}\right)^2=-\frac{11}{36}$$

Stop here: There are no real number solutions because no real number squared can equal $-\frac{11}{36}$.

9.3 SOLVING QUADRATIC EQUATIONS USING THE QUADRATIC FORMULA

Although completing the square can be used to solve any quadratic equation, it can be a long process. The quadratic formula can also be used to solve any quadratic equation, but you must memorize the formula to use it.

THE QUADRATIC FORMULA

The solutions to $ax^2 + bx + c = 0$, $a \neq 0$, are

$$x=\frac{-b\pm\sqrt{b^2-4ac}}{2a}$$

SOLVING QUADRATIC EQUATIONS USING THE QUADRATIC FORMULA

To solve a quadratic equation using the formula:

1. Write the equation in standard form: $ax^2 + bx + c = 0$. Multiply both sides by the LCD to clear fractions, if necessary.
2. Identify the values for a, b, and c.
3. Substitute the values for a, b, and c into the quadratic formula.
4. Simplify under the radical.
5. Simplify the radical, if possible.
6. Factor the numerator to reduce, if possible.

Exercise 9.4

Solve using the quadratic formula:

a) $x^2 - 5x + 4 = 0$

b) $x^2 + 4x = -1$

c) $4x^2 = -12x - 9$

d) $\frac{1}{6}x^2 + x = -\frac{5}{12}$

Solution 9.4

a) $x^2 - 5x + 4 = 0$ The equation is already in standard form.

$a = 1, b = -5, c = 4$ Identify the values for a, b, and c.

$x = \frac{-b \pm \sqrt{b^2 - 4ac}}{2a}$ Write the formula.

$x = \frac{-(-5) \pm \sqrt{(-5)^2 - 4(1)(4)}}{2(1)}$ Substitute the values for a, b, and c.

$x = \frac{5 \pm \sqrt{9}}{2}$ Simplify under the radical: $(-5)^2 - 4(1)(4) = 25 - 16 = 9$.

$x = \frac{5 \pm 3}{2}$ Simplify the radical.

$x = \frac{5+3}{2}$ or $x = \frac{5-3}{2}$ Solve for x.

$x = \frac{8}{2} = 4$ or $x = \frac{2}{2} = 1$

b) $x^2 + 4x = -1$

$x^2 + 4x + 1 = -1 + 1$ Add 1 to both sides to write the equation in standard form.

$x^2 + 4x + 1 = 0$ $ax^2 + bx + c = 0$.

$a = 1, b = 4, c = 1$ Identify the values for a, b, and c.

$x = \frac{-b \pm \sqrt{b^2 - 4ac}}{2a}$ Write the formula.

$x = \frac{-(4) \pm \sqrt{(4)^2 - 4(1)(1)}}{2(1)}$ Substitute the values for a, b, and c.

$x = \frac{-(4) \pm \sqrt{12}}{2}$ Simplify under the radical: $(4)^2 - 4(1)(1) = 16 - 4 = 12$.

$x = \frac{-(4) \pm 2\sqrt{3}}{2}$ Simplify the radical: $\sqrt{12} = \sqrt{4}\sqrt{3} = 2\sqrt{3}$.

$x = \frac{2(-2 \pm 1\sqrt{3})}{2}$ Factor the numerator.

$x = -2 \pm 1\sqrt{3}$ Reduce the fraction.

c) $4x^2 = -12x - 9$

$4x^2 + 12x + 9 = -12x - 9 + 12x + 9$

$4x^2 + 12x + 9 = 0$ Write the equation in standard form.

$a = 4,\ b = 12,\ c = 9$ Identify values for a, b, and c.

$x = \frac{-b \pm \sqrt{b^2 - 4ac}}{2a}$ Write the formula.

$x = \frac{-12 \pm \sqrt{(12)^2 - 4(4)(9)}}{2(4)}$ Substitute the values for a, b, and c.

$x = \frac{-12 \pm \sqrt{0}}{8}$ Simplify under the radical: $(12)^2 - 4(4)(9) = 144 - 144 = 0$.

$x = \frac{-12 \pm 0}{8}$ Simplify the radical: $\sqrt{0} = 0$.

$x = -\frac{12}{8} = -\frac{3}{2}$ Simplify the fraction.

d) $\frac{1}{6}x^2 + x = -\frac{5}{12}$

$\frac{1}{6}x^2 + x + \frac{5}{12} = -\frac{5}{12} + \frac{5}{12}$ Write the equation in standard form by adding $\frac{5}{12}$ to both sides.

$\frac{1}{6}x^2 + x + \frac{5}{12} = 0$

$12\left(\frac{1}{6}x^2 + x + \frac{5}{12}\right) = 12(0)$ Multiply by 12 to clear fractions.

$2x^2 + 12x + 5 = 0$

$a = 2, b = 12, c = 5$ Identify a, b, and c.

$x = \frac{-b \pm \sqrt{b^2 - 4ac}}{2a}$ Write the formula.

$x = \frac{-12 \pm \sqrt{(12)^2 - 4(2)(5)}}{2(2)}$ Substitute for a, b, and c.

$x = \frac{-12 \pm 2\sqrt{26}}{4}$ Simplify under the radical: $(12)^2 - 4(2)(5) = 144 - 40 = 104$.

$x = \frac{-12 \pm 2\sqrt{26}}{4}$ Simplify the radical: $\sqrt{104} = \sqrt{4 \cdot 26} = 2\sqrt{26}$.

$x = \frac{2(-6 \pm \sqrt{26})}{4}$ Factor the numerator.

$x = \frac{-6 \pm 1\sqrt{26}}{2}$ Reduce the fraction.

Stop here: You cannot reduce because –6 is a term, not a factor.

CHOOSING A METHOD TO SOLVE A QUADRATIC EQUATION

So far, you have seen several techniques for solving quadratic equations. When you are given a choice in the method used to solve a quadratic equation, you might consider the following pluses and minuses:

Method	Pluses	Minuses
Factoring	Usually fast	Doesn't always work
Completing the square	Always works: don't have to memorize a formula	Many steps
Quadratic formula	Always works: many calculators have the formula in memory	Must memorize formula

* If you like to factor, try factoring first.
* If you are good at memorizing formulas, use the quadratic formula.
* If you took the time to understand completing the square, go for it!

9.4 IMAGINARY AND COMPLEX NUMBERS

Previously, you saw that the square root of a negative number is not a real number. In this section, we make use of a new definition to explore the complex number system.

IMAGINARY NUMBERS

Square roots with negative radicands can be simplified using –1 as a factor. Numbers that are multiples of i are called **imaginary numbers,** where $i = \sqrt{-1}$ and $i^2 = -1$.

Exercise 9.5

Simplify:

a) $\sqrt{-7}$

b) $\sqrt{-4}$

c) $\sqrt{-18}$

Solution 9.5

a) $\sqrt{-7} = \sqrt{-1 \cdot 7}$ Factor.

$= \sqrt{-1}\sqrt{7} = i\sqrt{7}$ $\sqrt{-1} = i$

b) $\sqrt{-4} = \sqrt{-1 \cdot 4}$ Factor.

$= \sqrt{-1}\sqrt{4} = i \cdot 2$ $\sqrt{-1} = i$; $\sqrt{4} = 2$

$= 2i$ Generally, the number is written before the i.

c) $\sqrt{-18} = \sqrt{-1 \cdot 2 \cdot 9}$ Factor.

$= \sqrt{-1}\sqrt{2}\sqrt{9} = i \cdot \sqrt{2} \cdot 3$ $\sqrt{-1} = i$; $\sqrt{9} = 3$

$= 3i\sqrt{2}$

COMPLEX NUMBERS

A new set of numbers, the **complex numbers**, contains all the real numbers and the imaginary numbers. A complex number is in **standard form** when it is written as $a + bi$, where a and b are real numbers and $i = \sqrt{-1}$

Adding and Subtracting Complex Numbers

When adding and subtracting complex numbers, treat i like any other variable, except that i^2 can be simplified to -1.

Exercise 9.6

Combine the following complex numbers:

a) $(4 + 2i) + (6 - 7i)$

b) $(3 - 5i) - (2 - 9i)$

c) $(5 - 2i) - (4 - i) - (1 + 3i)$

Solution 9.6

a) $(4 + 2i) + (6 - 7i)$

$= (4 + 6) + (2 - 7)i$ — Combine similar terms. Treat i like a variable.

$= 10 - 5i$

b) $(3 - 5i) - (2 - 9i)$

$= (3 - 2) + (-5 + 9)i$ — Combine similar terms.

$= 1 + 4i$

c) $(5 - 2i) - (4 - i) - (1 + 3i)$

$= (5 - 4 - 1) + (-2 + 1 - 3)i$ — Combine similar terms.

$= 0 + (-4i) = -4i$

Multiplying Complex Numbers

Multiplication of complex numbers is similar to earlier multiplication with variables, except you must remember that i times i is i^2 which can be simplified to -1.

Exercise 9.7

Multiply:

a) $2i(4 - 3i)$

b) $(2 - 6i)(3 + 5i)$

Solution 9.7

a) $2i(4 - 3i) = 8i - 6i^2$ — Use the distributive property: $2i(4) - 2i(3i)$

$= 8i - 6(-1)$ — Use $i^2 = -1$.

$= 8i + 6 = 6 + 8i$ — Write in standard form.

b) $(2 - 6i)(3 + 5i)$

$2(3) + 2(5i) + (-6i)(3) + (-6i)(5i)$	Use FOIL.
$6 + 10i - 18i - 30i^2$	Combine similar terms: $(10 - 18)i = -8i$.
$= 6 - 8i - 30i^2$	
$= 6 - 8i - 30(-1)$	Use $i^2 = -1$.
$= 6 - 8i + 30$	
$= 36 - 8i$	Add similar terms.

Dividing Complex Numbers

If a complex number is written as a fraction with i in the denominator, we use a process similar to rationalizing the denominator for radicals to write that complex number in standard form. To write complex number quotients in standard form, do the following:

1. If the denominator is a monomial involving i, multiply by $\frac{-i}{-i}$ and simplify.
2. If the denominator is a binomial involving i, multiply the numerator and denominator by the conjugate of the denominator.

Recall that the conjugate of $a + b$ is $a - b$, and the conjugate of $a - b$ is $a + b$.

Exercise 9.8

Write each quotient in standard form:

a) $\frac{2+3i}{4i}$

b) $\frac{2-3i}{1+4i}$

c) $\frac{6}{2-3i}$

Solution 9.8

a) $\frac{2+3i}{4i} = \frac{2+3i}{4i} \cdot \frac{-i}{-i}$	Monomial denominator.
$= \frac{-2i-3i^2}{-4i^2}$	Distribute $-i$.
$= \frac{-2i-3(-1)}{-4(-1)}$	$i^2 = -1$.

$= \frac{-2i+3}{4}$ Simplify the numerator; simplify the denominator.

Note that some instructors will require that the answer be written in the following form:

$\frac{-2i}{4}+\frac{3}{4}=-\frac{i}{2}+\frac{3}{4}.$

b) $\frac{2-3i}{1+4i}=\frac{2-3i}{1+4i}\cdot\frac{1-4i}{1-4i}$ The conjugate of $1 + 4i$ is $1 - 4i$.

$= \frac{2-8i-3i+12i^2}{1-4i+4i-16i^2}$ Use FOIL to multiply the numerators and the denominators.

$= \frac{2-11i+12(-1)}{1-16(-1)}$ Replace i^2 with -1. Combine similar terms in the numerator and the denominator.

$= \frac{2-11i-12}{1+16}$ Combine similar terms in the numerator and denominator.

$= \frac{-10-11i}{17}$ or $-\frac{10}{17}-\frac{11i}{17}$

or $-\frac{10}{17}-\frac{11}{17}i$

c) $\frac{6}{2-3i}=\frac{6}{2-3i}\cdot\frac{2+3i}{2+3i}$ The conjugate of $2 - 3i$ is $2 + 3i$.

$= \frac{12+18i}{4+6i-6i-9i^2}$ Use the distributive property to multiply the numerators. Use FOIL to multiply the denominators.

$= \frac{12+18i}{4-9(-1)}$ Combine similar terms in the denominator: $+6i - 6i = 0$. Use $i^2 = -1$.

$= \frac{12+18i}{4+9}$ Simplify the denominator.

$= \frac{12+18i}{13}$ or $\frac{12}{13}+\frac{18}{13}i$

SOLVING QUADRATIC EQUATIONS WITH COMPLEX SOLUTIONS

You can now solve quadratic equations that have complex number solutions.

Exercise 9.9

Solve each quadratic equation using the method of your choice:

a) $(x - 3)^2 = -16$

b) $t^2 - 6t + 10 = 0$

c) $\frac{2}{5}y^2 = -\frac{1}{10}y - \frac{1}{2}$

d) $\left(x + \frac{1}{2}\right)^2 = -\frac{7}{4}$

e) $(x - 2)(x + 5) = -19$

Solution 9.9

a) $(x - 3)^2 = -16$ This is set up nicely for the square root property.

$\sqrt{(x-3)^2} = \pm\sqrt{-16}$ Take the square root of both sides.

$x - 3 = \pm 4i$ $\sqrt{-16} = \sqrt{-1 \cdot 16} = 4i$

$x - 3 + 3 = 3 \pm 4i$ Add 3 to both sides.

$x = 3 \pm 4i$ Solve for x.

b) $t^2 - 6t + 10 = 0$ Use the quadratic formula.

$a = 1, b = -6, c = 10$ Identify a, b, and c.

$t = \frac{-b \pm \sqrt{b^2 - 4ac}}{2a}$ Write the quadratic formula.

$t = \frac{-(-6) \pm \sqrt{(-6)^2 - 4(1)(10)}}{2(1)}$ Substitute a, b, and c.

$t = \frac{6 \pm \sqrt{-4}}{2}$ Simplify under the radical: $(-6)^2 - 4(1)(10) = 36 - 40 = -4$.

$t = \frac{6 \pm 2i}{2}$ Simplify the radical.

$t = \frac{2(3 \pm i)}{2}$ Factor the numerator.

$t = 3 \pm i$ Reduce the fraction.

c) $\frac{2}{5}y^2 = -\frac{1}{10}y - \frac{1}{2}$

$\frac{2}{5}y^2 + \frac{1}{10}y + \frac{1}{2} = -\frac{1}{10}y - \frac{1}{2} + \frac{1}{10}y + \frac{1}{2}$

$\frac{2}{5}y^2 + \frac{1}{10}y + \frac{1}{2} = 0$ Write the equation in $ax^2 + bx + c = 0$ form.

$10\left(\frac{2}{5}y^2 + \frac{1}{10}y + \frac{1}{2}\right) = (10)(0)$ Multiply by 10 to clear fractions.

$4y^2 + y + 5 = 0$ Identify $a = 4$, $b = 1$, and $c = 5$.

$y = \frac{-b \pm \sqrt{b^2 - 4ac}}{2a}$ Write the quadratic formula.

$y = \frac{-1 \pm \sqrt{(1)^2 - 4(4)(5)}}{2(4)}$ Substitute a, b, and c.

$y = \frac{-1 \pm \sqrt{-79}}{8}$ Simplify under the radical: $(1)^2 - 4(4)(5) = 1 - 80 = -79$.

$y = \frac{-1 \pm i\sqrt{79}}{8}$ Simplify the radical: $\sqrt{-79} = \sqrt{-1}\sqrt{79} = i\sqrt{79}$.

d) $\left(x + \frac{1}{2}\right)^2 = -\frac{7}{4}$ Use the square root property.

$\sqrt{\left(x + \frac{1}{2}\right)^2} = \pm\sqrt{-\frac{7}{4}}$ Take the square root of both sides.

$x + \frac{1}{2} = \pm\frac{i\sqrt{7}}{2}$ Simplify the radicals: $\sqrt{\frac{-7}{4}} = \frac{\sqrt{-1}\sqrt{7}}{\sqrt{4}} = \frac{i\sqrt{7}}{2}$.

$x + \frac{1}{2} - \frac{1}{2} = -\frac{1}{2} \pm \frac{i\sqrt{7}}{2}$ Subtract $\frac{1}{2}$ from both sides.

$x = \frac{-1 \pm i\sqrt{7}}{2}$

e) $(x - 2)(x + 5) = -19$ First write the equation in standard form.

$x^2 + 3x - 10 = -19$ FOIL the left side of the equation.

$x^2 + 3x - 10 + 19 = -19 + 19$ Add 19 to both sides of the equation.

$x^2 + 3x + 9 = 0$

$a = 1, b = 3, c = 9$ Identify a, b, and c to use the quadratic formula.

$x = \frac{-3 \pm \sqrt{(3)^2 - 4(1)(9)}}{2(1)}$ Substitute into the formula.

$x = \dfrac{-3 \pm 3i\sqrt{3}}{2}$ Simplify under the radical: $(3)^2 - 4(1)(9) = 9 - 36 = -27$.

Simplify the radical: $\sqrt{-27} = \sqrt{-1}\sqrt{9}\sqrt{3} = 3i\sqrt{3}$.

9.5 GRAPHING QUADRATIC EQUATIONS

The graph of a linear equation in two variables is a **straight line**. The graph of a quadratic equation in the form $y = ax^2 + bx + c$ is a **parabola**.

GRAPHING $y = ax^2 + bx + c$

To graph $y = ax^2 + bx + c$:

1. Find the y-intercept by substituting $x = 0$ into the equation and solving for y.
2. Find the x-intercept (if possible) by substituting $y = 0$ into the equation and solving for x.
3. Make a table of values, using a sufficient number of x values to determine the shape of the parabola (usually 4 or 5 values).
4. The parabola opens up if $a > 0$. The parabola opens down if $a < 0$.

Exercise 9.10

Graph:

a) $y = x^2$

b) $y = \dfrac{1}{4}x^2$

c) $y = -x^2 - 2x + 3$

Solution 9.10

a) $y = x^2$

If $x = 0$, $y = (0)^2$ Find the y-intercept.

$y = 0$

If $y = 0$, $(0) = x^2$ Find the x-intercept.

$\pm\sqrt{0} = \sqrt{x^2}$ Use the square root property to solve for x.

$0 = x$

$x = 0$

Make a table of values. List the intercept in the table:

x	y
0	0
1	1
2	4
–1	1
–2	4

Choose values for x, substitute each value into the original equation $y = x^2$, and solve for y.

If $x = 1$,	If $x = -1$,	If $x = 2$,	If $x = -2$,
$y = (1)^2$	$y = (-1)^2$	$y = (2)^2$	$y = (-2)^2$
$y = 1$	$y = 1$	$y = 4$	$y = 4$

To draw the graph, plot the points from your table:

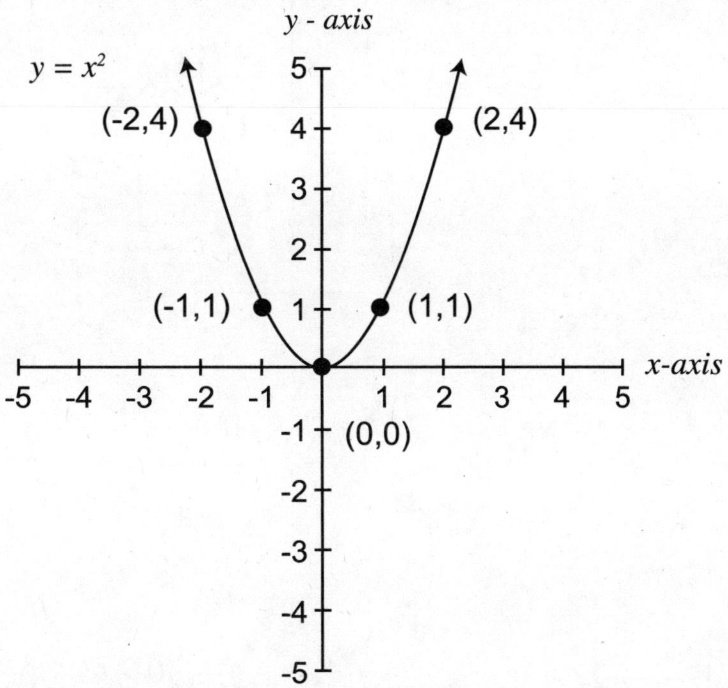

Notice that the parabola $y = x^2$ opens up. When the coefficient of x^2 is positive, the parabola will open up. (Here, the coefficient of $x^2 = 1x^2$ is 1.)

b) $y = \frac{1}{4}x^2$

If $x = 0, \quad y = \frac{1}{4}(0)^2$ — Find the y-intercept.

$y = 0$

If $y = 0, \; 0 = \frac{1}{4}x^2$ — Find the x-intercept(s).

$(0) = 4(\frac{1}{4}x^2)$ — Multiply by 4 to clear fractions.

$0 = x^2$

$\pm\sqrt{0} = \sqrt{x}$ — Take the square root of both sides.

$0 = x$

Make a table of values. List the intercept in the table:

x	y
0	0
–4	4
–2	1
2	1
4	4

If $x = -4$,	If $x = -2$,	If $x = 2$,	If $x = 4$,
$y = \frac{1}{4}(-4)^2$	$y = \frac{1}{4}(-2)^2$	$y = \frac{1}{4}(2)^2$	$y = \frac{1}{4}(4)^2$
$y = \frac{1}{4}(16)$	$y = \frac{1}{4}(4)$	$y = \frac{1}{4}(4)$	$y = \frac{1}{4}(16)$
$y = 4$	$y = 1$	$y = 1$	$y = 4$

To draw the graph, plot the points from your table:

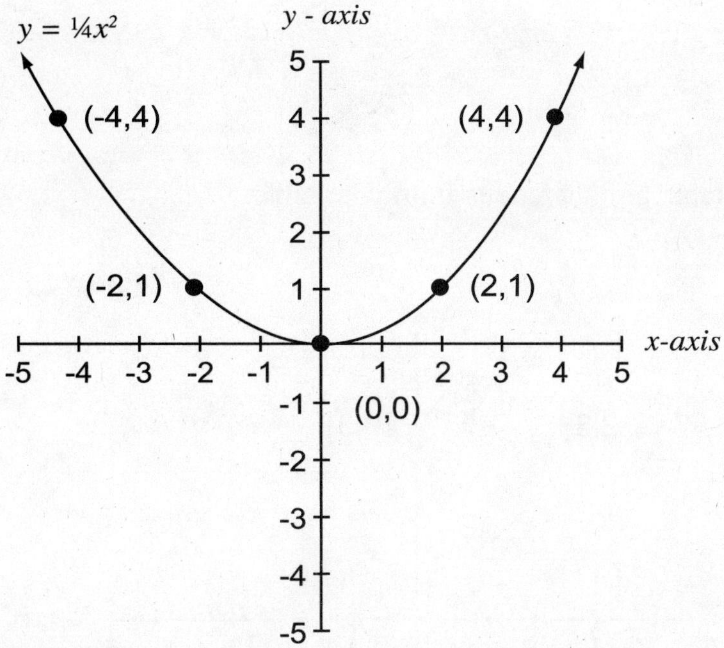

Notice that the parabola $y = \frac{1}{4}x^2$ opens up. When the coefficient of x^2 is positive, the parabola will open up. (Here, the coefficient of x^2 is $\frac{1}{4}$.)

c) $y = -x^2 - 2x + 3$

If $x = 0$, $y = -(0)^2 - 2(0) + 3$	Find the y-intercept.
$y = 3$	
If $y = 0$, $0 = -x^2 - 2x + 3$	Find the x-intercept.
$-1(0) = -1(x^2 - 2x + 3)$	Multiply by -1 to make factoring easier.
$0 = x^2 + 2x - 3$	This is a quadratic equation that factors.
$0 = (x + 3)(x - 1)$	Factor.
$x + 3 = 0$ or $x - 1 = 0$	Set each factor equal to 0 and solve for x.
$x = -3$ or $x = 1$	

Make a table of values. List the intercepts in the table:

x	y
0	3
–3	0
1	0
–1	4
–2	3

If $x = -1$,	If $x = -2$,
$y = -(-1)^2 - 2(-1) + 3$	$y = -(-2)^2 - 2(-2) + 3$
$y = -1 + 2 + 3$	$y = -4 + 4 + 3$
$y = 4$	$y = 3$

To draw the graph, plot the points from your table:

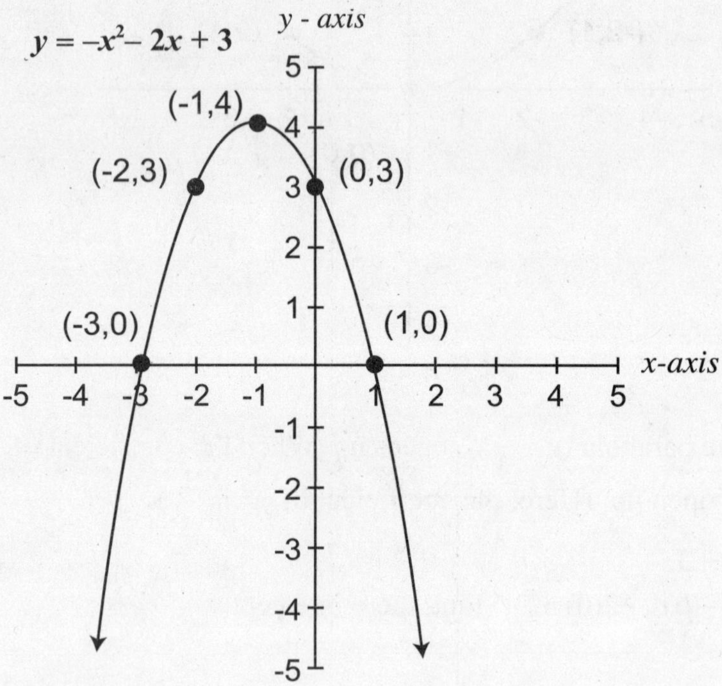

Notice that the parabola $y = -x^2 - 2x + 3$ opens down. When the coefficient of x^2 is negative, the parabola will open down. (Here, the coefficient of x^2 is -1.)

GRAPHING $y = a(x - h)^2 + k$

If the equation of the parabola is written in the form $\boldsymbol{y = a(x - h)^2 + k}$, you can use the **vertex** (the highest or lowest point) and the value of a to help you graph the equation.

To graph parabolas in the form $y = a(x - h)^2 + k$:

1. The vertex of the graph is the point (h, k).
2. If $a > 0$, the graph opens up from the vertex (h, k). If $a < 0$, the graph opens down from the vertex (h, k).
3. After you graph the vertex, select some values for x on both sides of the vertex to make a table of values.
4. Plot the points from your table and draw the parabola.

Exercise 9.11

Graph:

a) $y = (x - 1)^2 + 2$

b) $y = (x + 3)^2 - 1$

c) $y = -(x + 2)^2$

d) $y = x^2 - 3$

Solution 9.11

a) $y = (x - 1)^2 + 2$

Equation is in the form $y = a(x - h)^2 + k$, where $a = 1$, $h = 1$ and $k = 2$.

The vertex is (1, 2). The vertex is $(h, k) = (1, 2)$.

The graph opens up because $a = 1$. Because $a > 0$, the parabola opens up.

Make a table of values using x values on both sides of the vertex:

x	y
–1	6
0	3
1	2
2	3
3	6

If $x = -1$,

$$y = (x-1)^2 + 2$$
$$y = ((-1)-1)^2 + 2$$
$$y = (-2)^2 + 2$$
$$y = 4 + 2$$
$$y = 6$$

If $x = 0$,

$$y = (x-1)^2 + 2$$
$$y = ((0)-1)^2 + 2$$
$$y = (-1)^2 + 2$$
$$y = 1 + 2$$
$$y = 3$$

If $x = 2$,

$$y = (x-1)^2 + 2$$
$$y = ((2)-1)^2 + 2$$
$$y = (1)^2 + 2$$
$$y = 1 + 2$$
$$y = 3$$

If $x = 3$,

$$y = (x-1)^2 + 2$$
$$y = ((3)-1)^2 + 2$$
$$y = (2)^2 + 2$$
$$y = 4 + 2$$
$$y = 6$$

Plot the points from your table and draw the parabola:

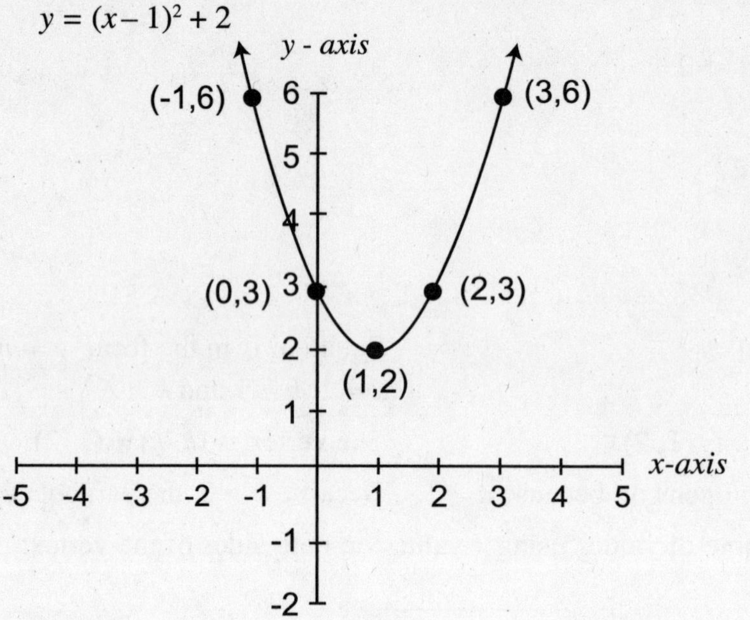

b) $y = (x + 3)^2 - 1$

$y = (x + 3)^2 + (-1)$	$-1 = +(-1) \quad +3 = -(-3)$
$y = (x - (-3))^2 + (-1)$	Rewrite in $y = a(x - h)^2 + k$ form. $a = 1$, $h = -3$, $k = -1$.
The vertex is $(-3, -1)$.	The vertex is $(h, k) = (-3, -1)$.

It may be easier to remember to change the sign of the number with x, and keep the sign of the number behind to find the vertex.

The graph opens up because $a = 1$. Because $a > 0$, the parabola opens up.

Make a table of values using x values on both sides of the vertex:

x	y
–5	3
–4	0
–3	–1
–2	0
–1	3

If $x=-5$,	If $x=-4$,	If $x=-2$,	If $x=-1$,
$y=(x+3)^2-1$	$y=(x+3)^2-1$	$y=(x+3)^2-1$	$y=(x+3)^2-1$
$y=((-5)+3)^2-1$	$y=((-4)+3)^2-1$	$y=((-2)+3)^2-1$	$y=((-1)+3)^2-1$
$y=(-2)^2-1$	$y=(-1)^2-1$	$y=(1)^2-1$	$y=(2)^2-1$
$y=4-1$	$y=1-1$	$y=1-1$	$y=4-1$
$y=3$	$y=0$	$y=3$	$y=3$

Plot the points from your table and draw the parabola:

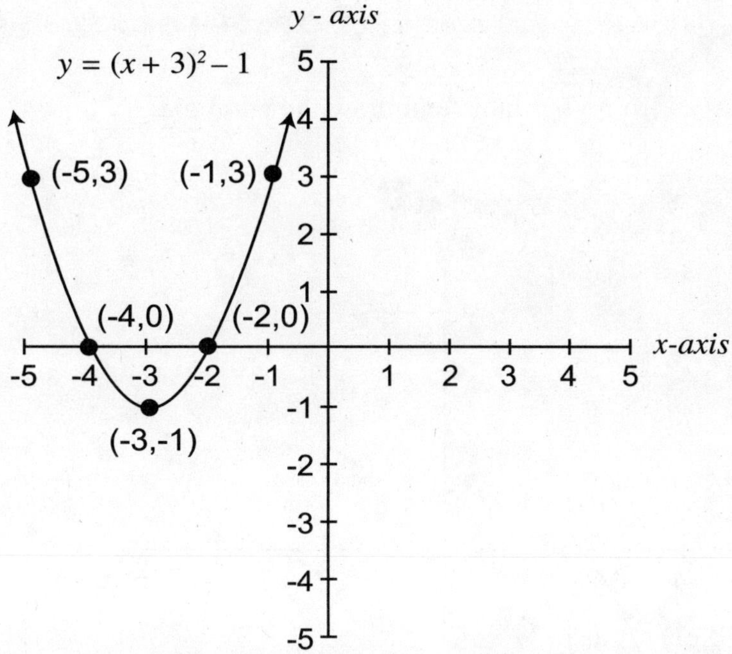

c) $y = -(x+2)^2$

$y = -1(x-(-2))^2 + 0$ — Rewrite in $y = a(x-h)^2 + k$ form. $a = -1$, $h = -2$, $k = 0$.

The vertex is $(-2, 0)$. — The vertex is $(h, k) = (-2, 0)$.

The graph opens down because $a = -1$. — Because $a < 0$, the parabola opens down.

Make a table of values using x values on both sides of the vertex:

x	y
–4	–4
–3	–1
–2	0
–1	–1
0	–4

If $x=-4$,	If $x=-3$,	If $x=-1$,	If $x=0$,
$y=-(x+2)^2$	$y=-(x+2)^2$	$y=-(x+2)^2$	$y=-(x+2)^2$
$y=-((-4)+2)^2$	$y=-((-3)+2)^2$	$y=-((-1)+2)^2$	$y=-((0)+2)^2$
$y=-(-2)^2$	$y=-(-1)^2$	$y=-(1)^2$	$y=-(2)^2$
$y=-(4)$	$y=-(1)$	$y=-(1)$	$y=-(4)$
$y=-4$	$y=-1$	$y=-1$	$y=-4$

Plot the points from your table and draw the parabola:

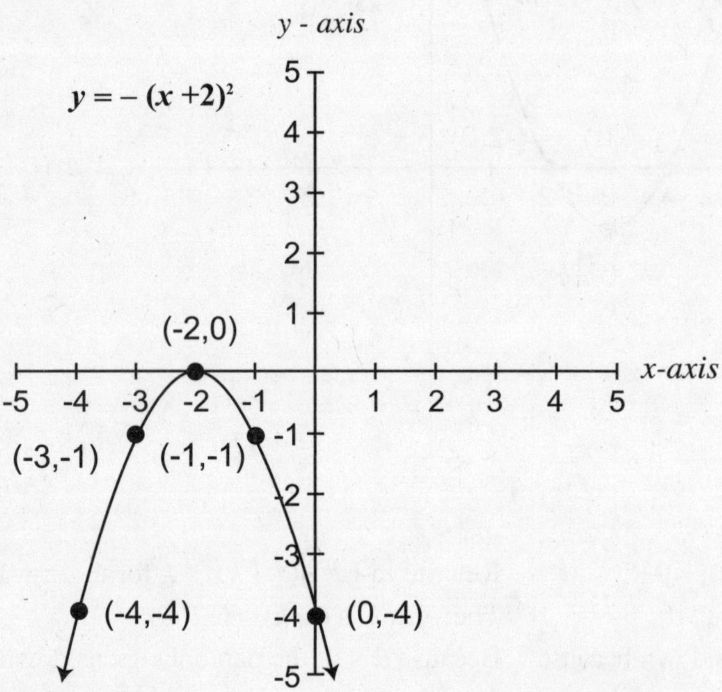

d) $y=x^2-3$

$y=1(x-0)^2+(-3)$ — Rewrite in $y=a(x-h)^2+k$ form. $a=1$, $h=0$, $k=-3$.

The vertex is $(0, -3)$. — The vertex is $(h, k)=(0, -3)$.

The graph opens up because $a=1$. — Because $a>0$, the parabola opens up.

Make a table of values using x values on both sides of the vertex:

x	y
-2	1
-1	-2
0	-3
1	-2
2	1

If $x=-2$,	If $x=-1$,	If $x=1$,	If $x=2$,
$y=x^2-3$	$y=x^2-3$	$y=x^2-3$	$y=x^2-3$
$y=(-2)^2-3$	$y=(-1)^2-3$	$y=(1)^2-3$	$y=(2)^2-3$
$y=4-3$	$y=1-3$	$y=1-3$	$y=4-3$
$y=1$	$y=-2$	$y=-2$	$y=1$

Plot the points from your table and draw the parabola:

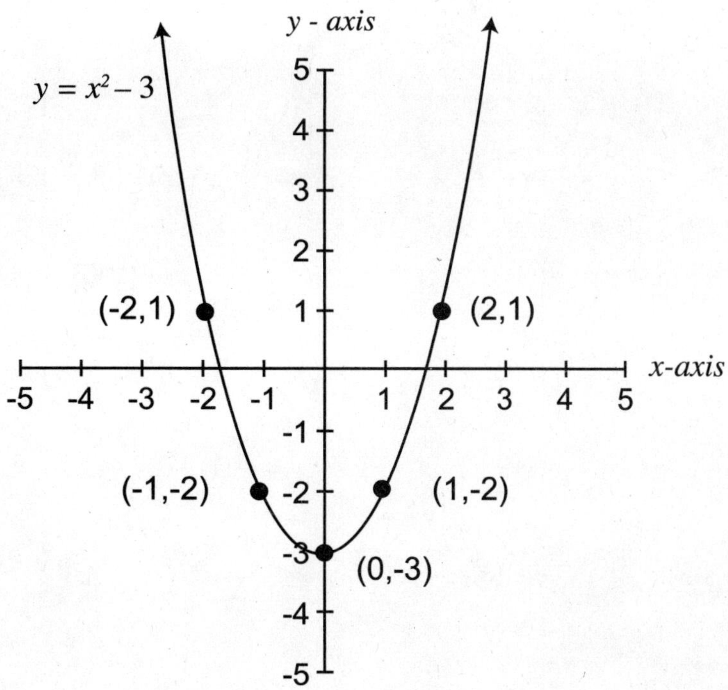

Exercise 9.12

Graph:

a) $y = \frac{1}{2}(x-2)^2 + 3$

b) $y = -4(x + 3)^2 + 1$

Solution 9.12

a) $y = \frac{1}{2}(x-2)^2 + 3$ Already in $y = a(x - h)^2 + k$ form.

$a = \frac{1}{2}, h = 2,$ and $k = 3.$

The vertex is (2, 3). The vertex is $(h, k) = (2, 3)$.

The graph opens up. Because $a > 0$, the graph opens up.

Make a table of values using x values on both sides of the vertex:

x	y
0	5
1	3.5
2	3
3	3.5
4	5

If $x = 0$,

$$y = \frac{1}{2}(x-2)^2 + 3$$
$$y = \frac{1}{2}((0)-2)^2 + 3$$
$$y = \frac{1}{2}(-2)^2 + 3$$
$$y = \frac{1}{2}(4) + 3$$
$$y = 2 + 3$$
$$y = 5$$

If $x = 1$,

$$y = \frac{1}{2}(x-2)^2 + 3$$
$$y = \frac{1}{2}((1)-2)^2 + 3$$
$$y = \frac{1}{2}(-1)^2 + 3$$
$$y = \frac{1}{2}(1) + 3$$
$$y = \frac{1}{2} + 3$$
$$y = 3.5$$

If $x = 3$,

$$y = \frac{1}{2}(x-2)^2 + 3$$
$$y = \frac{1}{2}((3)-2)^2 + 3$$
$$y = \frac{1}{2}(1)^2 + 3$$
$$y = \frac{1}{2}(1) + 3$$
$$y = \frac{1}{2} + 3$$
$$y = 3.5$$

If $x = 4$,

$$y = \frac{1}{2}(x-2)^2 + 3$$
$$y = \frac{1}{2}((4)-2)^2 + 3$$
$$y = \frac{1}{2}(2)^2 + 3$$
$$y = \frac{1}{2}(4) + 3$$
$$y = 2 + 3$$
$$y = 5$$

Plot the points from your table and draw the parabola:

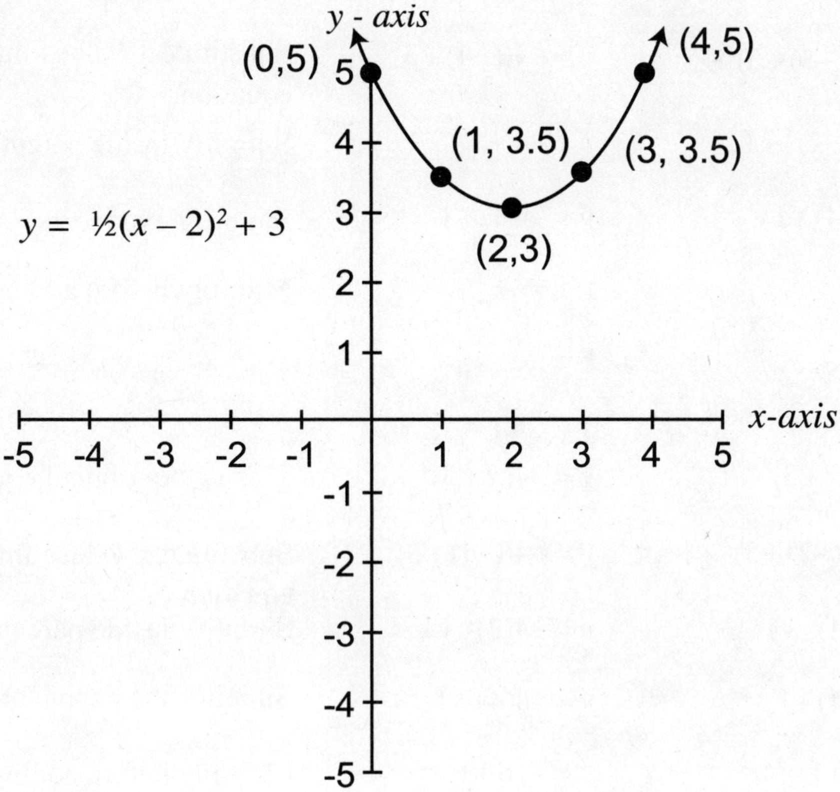

b) $y = -4(x + 3)^2 + 1$

$y = -4(x - (-3))^2 + 1$ Write in $y = a(x - h)^2 + k$ form. $a = -4$, $h = -3$, and $k = 1$.

The vertex is (–3, 1). The vertex is $(h, k) = (-3, 1)$.

The graph opens down. Because $a < 0$, the parabola opens down.

Make a table of values:

x	y
–5	–15
–4	–3
–3	1
–2	–3
–1	–15

If $x = -5$;	If $x = -4$	
$y = -4(x+3)^2 + 1$;	$y = -4(x+3)^2 + 1$	This is the original equation.
$y = -4((-5)+3)^2 + 1$;	$y = -4((-4)+3)^2 + 1$	Substitute x values into the original equation.
$y = -4(-2)^2 + 1$;	$y = -4(-1)^2 + 1$	Simplify inside parentheses first.
$y = -4(4) + 1$;	$y = -4(1) + 1$	Simplify exponents.
$y = -16 + 1$;	$y = -4 + 1$	Multiply before adding.
$y = -15$;	$y = -3$	

If $x = -2$;	If $x = -1$	
$y = -4(x+3)^2 + 1$;	$y = -4(x+3)^2 + 1$	This is the original equation.
$y = -4((-2)+3)^2 + 1$;	$y = -4((-1)+3)^2 + 1$	Substitute x values into the original equation.
$y = -4(1)^2 + 1$;	$y = -4(2)^2 + 1$	Simplify inside parentheses first.
$y = -4(1) + 1$;	$y = -4(4) + 1$	Simplify the exponents.
$y = -4 + 1$;	$y = -16 + 1$	Multiply before adding.
$y = -3$;	$y = -15$	

Plot the points from your table and draw the parabola:

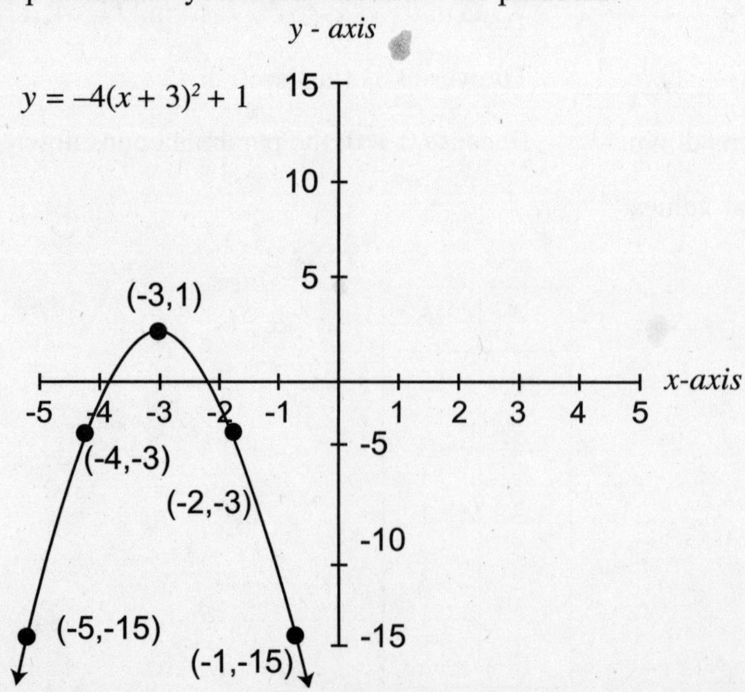

WRITING AN EQUATION IN $y = a(x - h)^2 + k$ FORM

You may need to complete the square to write the equation in $\boldsymbol{y = a(x - h)^2 + k}$ form. Study the following example:

Exercise 9.13

Find the vertex for the following:

a) $y = x^2 - 6x + 4$
b) $y = x^2 + 4x - 6$

Solution 9.13

a) $y = x^2 - 6x + 4$

$y - 4 = x^2 - 6x + 4 - 4$	Isolate $x^2 - 6x$ by subtracting 4 from both sides.
$y - 4 = x^2 - 6x$	$\left(\frac{1}{2} \cdot -6\right)^2 = (-3)^2 = 9$.
$y - 4 + 9 = x^2 - 6x + 9$	Add 9 to both sides.
$y + 5 = x^2 - 6x + 9$	Factor $x^2 - 6x + 9$.
$y + 5 = (x - 3)^2$	
$y + 5 - 5 = (x - 3)^2 - 5$	Subtract 5 from both sides.
$y = (x - 3)^2 - 5$	Write in $y = a(x - h)^2 + k$ form.
$y = (x - 3)^2 + (-5)$	
Vertex is $(3, -5)$	The vertex is $(h, k) = (3, -5)$.

b) $y = x^2 + 4x - 6$

$y + 6 = x^2 + 4x - 6 + 6$	Isolate $x^2 + 4x$ by adding 6 to both sides of the equation.
$y + 6 = x^2 + 4x$	$\left(\frac{1}{2} \cdot 4\right)^2 = (2)^2 = 4$.
$y + 6 + 4 = x^2 + 4x + 4$	Add 4 to both sides.
$y + 10 = (x + 2)^2$	Factor $x^2 + 4x + 4$.
$y + 10 - 10 = (x + 2)^2 - 10$	Subtract 10 from both sides to write in $y = a(x - h)^2 + k$ form.
$y = (x + 2)^2 - 10$	
$y = (x - (-2))^2 + (-10)$	
Vertex is $(-2, -10)$	The vertex is $(h, k) = (-2, -10)$.

TEST YOURSELF

1) Solve:

(a) $x^2 = 100$

(b) $(x + 3)^2 = 36$

(c) $25m^2 - 6 = 42$

(d) $(x - 2)^2 = -25$

2) Solve by completing the square:

(a) $x^2 + 2x - 15 = 0$

(b) $x^2 + x - 12 = 0$

(c) $x^2 - 10x - 2 = 0$

3) Solve by completing the square:

(a) $2x^2 + 12x - 22 = 0$

(b) $3x^2 - 15x - 3 = 0$

(c) $x^2 + 10x + 57 = 0$

4) Solve using the quadratic formula:

(a) $x^2 + 4x - 12 = 0$

(b) $x^2 - 8x - 12 = 0$

(c) $x^2 = 3x + 9$

(d) $\frac{1}{5}x^2 - x = 1$

5) Simplify:

(a) $\sqrt{-11}$

(b) $\sqrt{-128}$

(c) $\sqrt{-49}$

6) Combine the following complex numbers:

(a) $(5 + 3i) + (2 - 11i)$

(b) $(2 - 7i) - (8 - i)$

(c) $(3 + 2i) - (4 - 5i) - (6 + i)$

7) Multiply:

(a) $3i(2 + 5i)$

(b) $(4 - 3i)(2 + 5i)$

(c) $(3 - 2i)(3 + 2i)$

8) Write each quotient in standard form:

(a) $\frac{5+2i}{3i}$

(b) $\frac{1-4i}{3-2i}$

(c) $\frac{4}{2+i}$

9) Solve each quadratic equation:

(a) $(x + 2)^2 = -25$

(b) $x^2 - 8x + 43 = 0$

(c) $\frac{1}{6}x^2 = x - \frac{25}{6}$

10) Graph:

(a) $y = (x + 1)^2$

(b) $y = (x - 2)^2 + 1$

(c) $y = x^2 + 3$

TEST YOURSELF ANSWERS

1) (a) ± 10

(b) $-9, 3$

(c) $\pm \frac{4\sqrt{3}}{5}$

(d) No real solutions

2) (a) $-5, 3$

(b) $-4, 3$

(c) $5 \pm 3\sqrt{3}$

3) (a) $-3 \pm 2\sqrt{5}$

(b) $\frac{5 \pm \sqrt{29}}{2}$

(c) No real solutions

4) (a) $-6, 2$

(b) $4 \pm 2\sqrt{7}$

(c) $\frac{3 \pm 3\sqrt{5}}{2}$

(d) $\frac{5 \pm 3\sqrt{5}}{2}$

5) (a) $i\sqrt{11}$

(b) $8i\sqrt{2}$

(c) $7i$

6) (a) $7 - 8i$

(b) $-6 - 6i$

(c) $-7 + 6i$

7) (a) $-15 + 6i$

(b) $23 + 14i$

(c) 13

8) (a) $\frac{2}{3} - \frac{5}{3}i$

(b) $\frac{11}{13} - \frac{10}{13}i$

(c) $\frac{8}{5} - \frac{4}{5}i$

9) (a) $-2 \pm 5i$

(b) $4 \pm 3i\sqrt{3}$

(c) $3 \pm 4i$

10) (a)

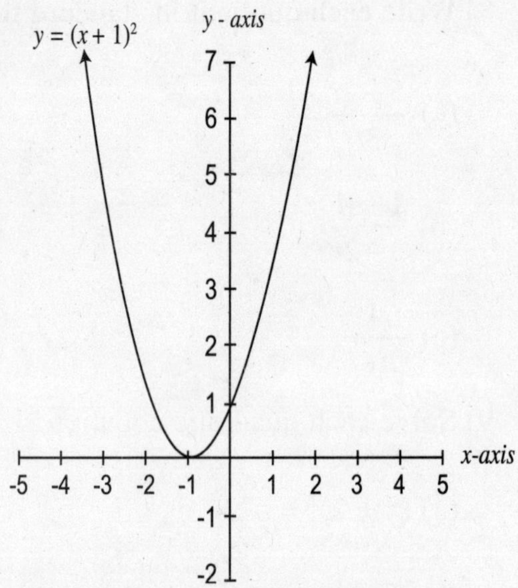

(b)

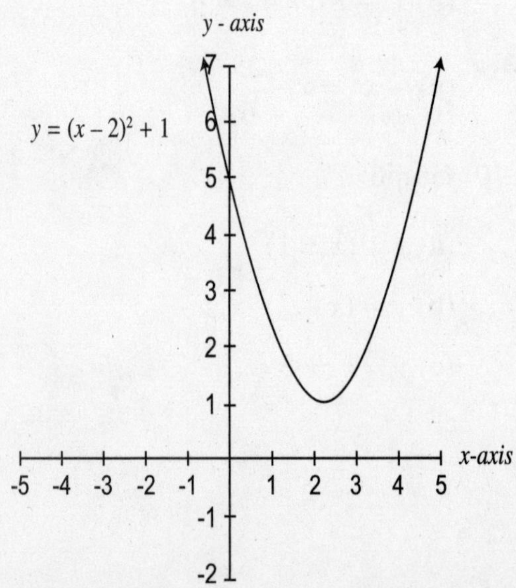

(c)

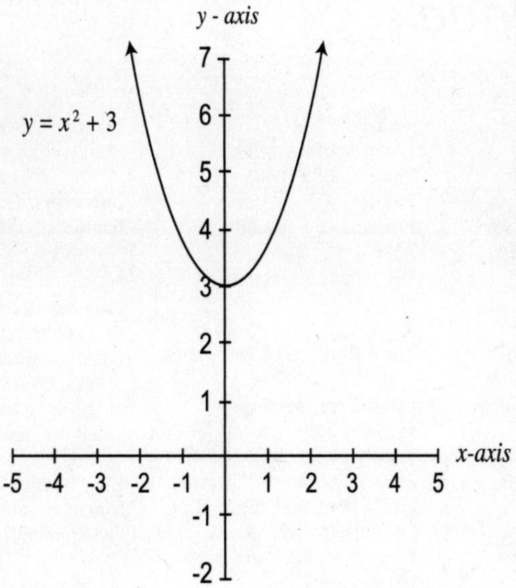

Index

Collins *College* OUTLINES

Fully Revised and Updated

Written by professors, teachers, and experts in various fields, the titles in the Collins College Outlines series provide students with a fast, easy, and simplified approach to the curricula of important introductory courses. Selected titles also provide perfect preparation for AP exams. Each title contains a full index and a "Test Yourself" section with full explanations for each chapter.

BASIC MATHEMATICS
Lawrence A. Trivieri
ISBN 0-06-088146-1 (paperback)

INTRODUCTION TO CALCULUS
Joan Van Glabek
ISBN 0-06-088150-X (paperback)

INTRODUCTION TO AMERICAN GOVERNMENT*
Larry Elowitz
ISBN 0-06-088151-8 (paperback)

INTRODUCTION TO PSYCHOLOGY*
Joseph Johnson and Ann L. Weber
ISBN 0-06-088152-6 (paperback)

MODERN EUROPEAN HISTORY
John R. Barber
ISBN 0-06-088153-4 (paperback)

ORGANIC CHEMISTRY
Michael Smith
ISBN 0-06-088154-2 (paperback)

UNITED STATES HISTORY TO 1877*
Light Cummins and Arnold M. Rice
ISBN 0-06-088159-3 (paperback)

WESTERN CIVILIZATION TO 1500
John Chuchiak and Walter Kirchner
ISBN 0-06-088162-3 (paperback)

ABNORMAL PSYCHOLOGY
Sarah Sifers
ISBN 0-06-088145-3 (paperback)

UNITED STATES HISTORY FROM 1865*
John Baick and Arnold M. Rice
ISBN 0-06-088158-5 (paperback)

ELEMENTARY ALGEBRA
Joan Van Glabek
ISBN 0-06-088148-8 (paperback)

SPANISH GRAMMAR
Ana Fairchild and Juan Mendez
ISBN 0-06-088157-7 (paperback)

***PERFECT FOR AP REVIEW**

Available wherever books are sold, or call 1-800-331-3761 to order.